Manual of Contract Documents for Highway Works

a user's guide and commentary

Volume 1

Bill Money and Geoff Hodgson

Published by Thomas Telford Services Ltd, Thomas Telford House, 1 Heron Quay, London E14 4JD

First published 1992

A catalogue record for this book is available from the British Library

ISBN for set of 2 volumes: 0 7277 1909 2
ISBN for volume 1: 0 7277 1911 4

The information contained in this book is intended for use as a general statement and guide only. The authors and publishers cannot accept any liability for any loss or damage which may be suffered by any person as a result of the use in any way of the information contained herein. Any person drafting contracts, specifications or other documents based on the Manual of Contract Doucments for highway works must in all cases use those documents and take appropriate professional advice on the matters referred to in this book. Users themselves are solely responsible for ensuring that any wording taken from these notes is consistent with and appropriate to the remainder of their material.

Typeset in Great Britain by MHL Typesetting Ltd, Coventry

Printed and bound in Great Britain by Redwood Press Limited, Melksham, Wiltshire

Contents

Preface

In 1988 Steve Arnold (Chairman of CTA Services Ltd) suggested that Bill Money should consider researching the then relatively new Department of Transport Specification for Highway Works with a view to running courses on the subject. At the time the Brown Book and its associated Method of Measurement for Highway Works were just starting to come into general use and CTA had been requested by many organisations to run courses on the new documents. Bill's background of running a contracts section in a county council's highway department had involved him in many of the intricacies of the old Blue Book and its Method of Measurement and he therefore decided to take up Steve's suggestion.

The resultant courses, which covered both the general aspects of the documents and the specialised aspects of claims, measurement and earthworks, proved to be extremely popular and were run throughout the UK for consultants, local authorities and contractors. In particular the linkage of the measurement rules to the specification requirements was clearly of benefit and often highlighted areas of misunderstanding in dealing with the new documents.

It became increasingly apparent from the course feedback that there was a need for a book bringing together the various documents and offering advice on the areas where many mistakes were occuring but Bill's other commitments to running contractual courses and to his consultancy work prevented it being taken any further.

The popularity of the courses was also producing problems both for Bill in making time available and for Barbara French of CTA Services who was responsible for marketing a product where demand at times far exceeded supply. Consequently it was a great relief when in 1991 Geoff Hodgson, a lecturer in Quantity Surveying at Birmingham Polytechnic, agreed to do some of the courses using Bill's original research and notes.

By this time the Seventh Edition was in the offing and an approach to John Kerman head of Engineering Policy and Programming Division at the DTp. resulted in the generous provision of advance information on the draft documents to Bill and Geoff. This enabled research to commence with a view to organising training courses to coincide with the publication of the new Manual of Contract Documents.

The idea of a book resurfaced at this stage as it seemed an appropriate time to produce such a document which would build on the experience gained on the Sixth Edition courses and highlight the changes introduced in the Seventh Edition with obvious benefits to users of the new Manual.

The authors believe this volume and its companion will save a great deal of time in comparing the new and old documents and in the background reading necessary for those engineers and quantity surveyors involved in preparing contract documents, tendering, supervising, constructing and measuring the work. It brings together in an accessible form the requirements of the Specification, advice from the Notes for Guidance and other background documentation, and links these to the rules of measurement. Whilst it highlights the changes to the documents brought about by the Seventh Edition (the changes are shown in the clause notes by italics) it comprehensively covers all aspects of the documents whether these have changed or not. It also draws to the user's attention those areas of difficulty where problems are likely to occur and where there are anomalies or inconsistencies in the documents.

It is not intended in any way to replace a proper use of the documents themselves but should ensure that the user is directed to those areas of change and where particular care needs to be taken in interpretation.

Without the generous assistance from the DTp. in making available the draft publication documents and background information it would not have been possible to have produced this book in the very short timescale available.

The authors would also like to acknowledge the kind permission of HMSO and the Department of Transport to reproduce parts of the published documents in this volume.

Geoff Hodgson would like to acknowledge Bill Money for all the technical advice and guidance he provided in the writing of both volumes of this book. Geoff would also like to thank the School of Quantity Surveying at the University of Central England in Birmingham, for their cooperation in allowing time for the writing of these volumes.

W. Money and G.J. Hodgson, July 1992

Introduction to the DTp. Specification for Highway Works (Seventh Edition)

The DTp. Specification for Highway Works (SHW) has become a standard component of UK highway contracts since its introduction in 1951, and has developed through the six previous revisions which have taken place on average every six or seven years. In order to benefit from research and experience in road building and road and bridge maintenance the Specification has evolved and inevitably has become a more lengthy and complex document in the process.

As a crucial contract document it underpins the system of competitive tendering and this has tended to encourage a degree of conservatism with a tendency to constrain the Contractor's freedom of choice, sometimes by specifying by product name rather than by end performance. In recent years pressure has grown to relax some of these constraints so as to encourage the development of innovatory procedures and new materials and also to remove potential barriers to trade which may prevent the use of otherwise acceptable products from other EC member states. It is this second factor, and the imminence of the Single European Market which has been the driving force behind the Seventh Edition of the SHW.

Barriers to Trade

European legislation and in particular Notification Directive 83/189 requires that the Specification for Highway Works and other documents have to be notified to the European Commission and it was recognised that these would need to be updated and potential 'barriers to trade' removed prior to submission (see the 100 Series notes for discussion of 'barriers to trade'). Consequently a commission was let to W.S. Atkins with the objective:

> 'To amend the 6th Edition of the SHW to take account of European legislation, subsequent amendments and other changes and republish. Revise associated documents consequent to changes in SHW. Advise on consequential changes needed to the Department's Model Contract Document'.

Notification Directive 83/189 requires 'technical regulations' to be notified to the European Commission who, after a three-month notification period, issue objections or 'detailed opinions' on the submitted information. The definition of technical regulations includes technical specifications and their associated administrative procedures.

The documents were first updated to take account of the various technical amendments and then examined to identify and remove potential barriers to trade. The SHW, Notes for Guidance and Highway Construction Details were then submitted to the EC and consultation with various UK bodies was carried out during the notification period (FCEC, ICE, County Surveyors Society, manufacturers' trade associations).

The detailed opinions received from the EC were then examined and responded to, often with a lengthy process of consultation, to produce fully Europeanised documents. The detailed opinions ranged from various technical aspects to the overall approach adopted to deal with common standards.

The consequential effects of these changes also needed to be assessed in relation to the MMHW and Library of Standard Item Descriptions and these documents were amended accordingly after consultation with FCEC.

Ease of Use

As well as the necessary changes because of the EC dimension it was also recognised that there was a need to improve the user-friendliness of contract documentation, given the increasing complexity of modern contracts. The Specification should be easy to use by compilers of contract documents, and should lead to the production of documents which are easy to interpret by tenderers.

The Specification should form part of a controlled system of standard documentation, reflecting the move towards use of quality assurance procedures in design offices and on site. It should be easy to update, so as to allow for the introduction of new developments and techniques at more frequent intervals than the six or seven years which has become the norm.

In recognition of these aims it was decided that a loose-leaf manual comprising all the documents necessary to draw up a trunk road contract would be the preferred solution. With the aim of improving the understanding of the MMHW it was also decided to reintroduce the Notes for Guidance to the MMHW as experience has shown that this document is poorly understood and often mis-used by those measuring and supervising contracts.

Four volumes of the Manual are published by HMSO and will comprise:

Volume 1 Specification for Highway Works
Volume 2 Notes for Guidance on the SHW
Volume 3 Highway Construction Details
Volume 4 Bills of Quantities for Highway Works (Method of Measurement for Highway Works, Notes for Guidance on the MMHW, and the Library of Standard Item Descriptions)

In addition a preliminary volume will be issued by the Department of Transport as:

Volume 0 Model Contract Document for Major Works with the standards which formally require the introduction of the HMSO documents on trunk road contracts, and associated advice notes.

Complementing the Manual of Contract Documents for Highway Works is a sister Design Manual for Roads and Bridges which draws together into a coherent framework the various engineering design standards and advice notes.

With a loose-leaf format there is a need to maintain rigorous control of amendments and to ensure that the correct version of the Specification is used and can be referred to without ambiguity for each contract. It is therefore the intention that national amendments will be issued annually together with a schedule of page numbers and publication dates with the schedule being bound into the contract documents for a specific scheme. This procedure reflects the DTp.'s philosophy of accepting that they cannot introduce all the desirable changes to the Specification immediately and therefore wish to prepare the ground well in advance of the first annual amendment.

Implementation

Various standards and Advice Notes cover the implementation of the SHW and Notes for Guidance (SD 1/92), Highway Construction Details (SD 2/92), Bills of Quantities (SD 3/92), the adoption of proprietary structures (SD 4/92), type approved or registered products (SA1/92), assessing equivalence (SA 2/92) and testing (SA 3/92).

In terms of timing the DTp. consider that it should not take long to transfer preparation of contract documents from the Sixth to the Seventh Edition and expect that, unless there are overriding reasons for continuing with the Sixth Edition, contracts let after 31 December 1992 will be on the new documents.

Initially the Specification will be introduced using the ICE Conditions of Contract (Fifth Edition) although it is anticipated that the ICE Sixth Edition may be introduced, possibly with amendment, during 1993. Where necessary the ICE Conditions are being amended in the Model Contract Document to suit the changes in the Specification (e.g. in respect of Contractor's design).

100 SERIES Preliminaries

Introduction

The 100 Series contains the most fundamental changes from the previous Sixth Edition with major new concepts contained in Clauses 104 and 106, major changes in Clause 105, new clauses 107, 123 and 124 and important changes to 117, 118 and 120. This introduction serves to give an insight into the reasons for the changes and the notes on the individual clauses will provide detailed information on the changes in the documents themselves and should be read in conjunction with these background notes.

In reading the detailed clause notes the wording in italics signifies the areas of change compared with the Sixth Edition.

Whilst the changes to 105 have been made because of the need to overcome problems identified on previous contracts the motivating force behind the fundamental changes to Clauses 104 and 106 has been EC legislation and particularly the need to avoid 'barriers to trade'.

Article 30 of the Treaty of Rome states that:

> 'Quantitative restrictions on imports and all measures having equivalent effect shall, without prejudice to the following provisions, be prohibited between member states'

The 'following provisions' dealt with restrictions in force at the instigation of the treaty and transitional arrangements.

Following court cases in Europe it has been established that qualititative restrictions also contravene Article 30. An infringment of Article 30 is a 'barrier to trade' (see Boxes 1 and 2 for examples) and although provision is made for certain circumstances where restrictions may be permissible, these cannot be used to disguise restrictions or discrimination.

In cases considered by the European Court of Justice it is the words 'all measures having equivalent effect' that have been elaborated to establish what constitutes a 'barrier to trade'. Thus clauses specifying 'all steel reinforcement to be of British origin' are outlawed, as are clauses requiring tests to be carried out by a British testing consultant. Similarly, the use of trade names or specifications which point only to a particular product are no longer permissible.

The following principles have been developed from Article 30 and are put forward as guidelines to assist those involved in drawing up public procurement specifications and contracts:

(a) **Transparency** — all requirements must be open, clear and readily acceptable to those who need to know — there must be no hidden systems.
(b) **Reciprocity** — public procurement organisations must recognise other national standards and properly undertaken tests in other member states where these are equivalent to those specified.
(c) **Proportionality** — the means of achieving an objective must be commensurate with the end being pursued. In contracts this means a product should not be overspecified nor should tests be too stringent.

The Construction Products Directive (CPD) is now in force which makes it unlawful for public procurers to impede the use of products complying

The following are some circumstances which could give rise to the creation of a barrier to trade. It must be appreciated that the actual circumstances are important in any case and there may be rare occasions when some of these actions are justified.

— specifying proprietary products where suitable standards exist

— specifying proprietary products without acknowledging that equivalents will be acceptable

— requiring tests to be carried out at a particular place as a condition of acceptance of a product

— specifications which give an advantage to certain products

— refusing to accept results of test properly undertaken in another member state

— applying different criteria to products from outside the UK

— for a government body or department to require evidence that a product complies with a regulation or an approval scheme to be in English

— specifying unnecessary product characteristics such as stainless steel pipes for ordinary drains

— specifying the country or place of origin of a product

— unjustified requirements for Departmental approval of a product

Box 1 Examples of Barriers to Trade

The question of language is particularly difficult — whilst a 'customer' can reasonably demand information in his own language it is thought likely that a government or major public body should have resources to translate submissions in any of the nine official EC languages.

The position with a government or agency contract is uncertain but it is assumed that the position of Engineer would be analogous to that of an individual 'customer' and SHW Clause 104 requires an English translation.

Box 2 Language

with the provisions of the Directive. Levels of safety, suitability and fitness for purpose (within the Essential Requirements of the CPD) must not exceed what is necessary to ensure that the works comply with them. The Essential Requirements are written in general terms and the DTp. do not consider it likely that specifications in department contracts could be held to be inconsistent.

The Seventh Edition of the SHW has been compiled with these principles in mind resulting in major changes from the Sixth Edition to ensure compliance. It is now fully 'Europeanised' having been through the Notification procedure.

Clause 104 introduces the principle of reciprocity into DTp. contracts and accepts the principle of 'equivalent' work, goods and materials. The principle of equivalence applies to standards, QA schemes, Agrément certificates or approvals of any member state of the EC.

The assessment of equivalence may be undertaken within or outside a specific contract. Within a specific contract the Engineer will normally have to decide if proposed equivalent goods and materials (PEQ) are equivalent. Outside specific contracts it will be for the Overseeing Department to assess equivalence in response to manufacturer's and supplier's requests (see also Box 8).

Equivalence does not mean that the proposed products must be identical but that they offer similar levels of safety, suitability and fitness for purpose. It should be assumed that standards will have to be adopted if they achieve or provide evidence of equivalent performance on behalf of the product concerned. Tests undertaken by proper testing houses in other member states must be recognised although acceptance of a testing house will require satisfactory evidence of its technical and professional competence and independence.

Whilst Clause 104 addresses the principle of equivalence it is also neccesary to avoid 'barriers to trade' created by the specifying of proprietary materials and this has been done through the introduction of 'Contractor's design' in Clause 106.

Clause 106 — *Design of Permanent Works by the Contractor* is essentially a new clause replacing the previous Alternative Specified Materials and introduces Contractor's design of structures, structural elements or other features where proprietary materials are the norm or where they may be a suitable alternative. Thus the specifier does not stipulate a particular proprietary material (which would be a barrier to trade) but leaves the choice and design to the Contractor. Such a procedure should also encourage the use of proprietary products in appropriate circumstances where they may offer better value for money.

In the case of structures, previous practice for including a proprietary product into a contract has been either

> to prepare a general arrangement drawing, based on a particular proprietary structure, and with 'or equivalent' added; or
>
> to prepare a fully detailed design in non-proprietary materials but be prepared to accept an alternative, proposed by the Contractor, based on a proprietary product.

Neither of these approaches satisfies the requirements to avoid barriers to trade and to encourage the use of these products.

Where a proprietary material can be identified from the contract that particular material is put at an advantage and tenderers will often consider the effort of preparing and pricing an alternative, in addition to the Engineer's design, a considerable deterrent. They can never be certain as to what might be acceptable and the additional design and technical approval procedures are quite onerous.

To overcome these objections a new approach has been taken with a new Standard SD 4/92 'Procedures for Adoption of Proprietary Structures' being published. The structures to be designed by the Contractor will be listed in Appendix 1/10(A) — or where a choice of designs is offered in Appendix 1/10B — and the contractual requirement will be an outline specification (O/AIP) which leaves the Contractor free to choose a product that meets the criteria subject to checking by the Engineer and Technical Approval (see Boxes 3 and 16).

Box 3 Adoption of Proprietary Manufactured Structures

The basic requirements of the new approach are:

— The Engineer prepares an outline specification for the structure with all the information necessary for a final design, but without any reference to a particular product.

— This Outline Approval in Principle form (O/AIP) should leave as much choice as possible to the tenderers

— The Engineer obtains Technical Approval to the O/AIP

— The tenderers price on a lump sum basis for any structure of their choice which meets these requirements

— All proprietary manufactured structures require specific technical approval to BD2/89

— After award of the Contract the Contractor completes the detailed design and submits it for technical approval through the Engineer

— The Engineer checks and forwards to TAA for Technical Approval

— Following technical approval the Engineer accepts the structure and the Contract proceeds as normal

In a situation where the Engineer considers that a non-proprietary design and a proprietary system are more or less equal in terms of cost and performance then both possibilities are run in parallel but the tenderer only prices his choice.

The MMHW 'Preparation of Bill of Quantities' paras 7 and 8 sets out the arrangements of the Bill of Quantities to cover these situations.

Structures Designed by the Contractor

16 In respect of each priced Bill for a structure designed by the Contractor, the Contractor shall prepare a priced schedule of quantities. This priced schedule shall be prepared in accordance with the relevant Chapters and Series of the Method of Measurement and submitted to the Engineer for his acceptance within 7 working days of the date of the Engineer's approval in writing of the Contractor's design. Reference to such structures and associated works within Chapters and Series other than Series 2500, are included to enable the Contractor to prepare the priced schedule of quantities and to enable the Engineer to prepare a Bill of Quantities where a specifically designed option is included for such a structure.

The quantities, rates and prices in the priced schedule of quantities shall in each case, when extended and totalled, give the amount entered in the priced Bill of Quantities against the item for the relevant structure designed by the Contractor.

The priced schedule of quantities shall only be used for the valuation of monthly statements and for the valuation in accordance with Clause 52 of the Conditions of Contract of variations ordered by the Engineer in connection with structures designed by the Contractor.

Unless specifically stated to the contrary in the Contract the measurement of the Works affected by the incorporation of the Contractor's design shall be based on the Tender documents and not on the Works as amended and completed to incorporate the Contractor's design.

The parts of the Works included by the Contractor in the priced schedule of quantities shall include all the parts of the Works within the Designated Outline except those designed and scheduled by the Engineer as not included.

Earthworks within the Designated Outlines shall not be included by the Engineer in the earthworks schedules.

Box 4 Preambles to Bill of Quantities

The Contractor's tender price is based on his choice of structure and the lump sum must cover all the work (with specified exceptions) within a carefully designated and fully enclosed part of the site (the designated outline).

The MMHW sets out the rules for measurement in General Directions 15 and 16 of the Preambles to Bill of Quantities (see Box 4) and a separate series (2500) of the MMHW has been introduced to deal with such structures.

Tenderers will need to judge for themselves the amount of detailed design needed at the tender stage (where appropriate the Contractor may propose a manufacturer's design) and their lump sum price will be based on this. It will be deemed to include any necessary revisions or amendments to design as part of the technical approval process and to allow for any consequential effects outside the designated outline (see Box 18).

The successful tenderer will complete the detailed design after the Contract is awarded and provide a schedule of quantities which must total to the lump sum.

It is obvious that with such a system there will be considerable changes required in the preparation of drawings and contract documents, in tendering and in measurement and supervision. Both the Contractor and the Engineer will have to take great care if problems are to be avoided:

> The O/AIP must allow as much choice as possible without compromising the Engineer's design requirements.
>
> The designated outline must be clearly defined, fully enclosing and large enough for any possible options — including any special backfilling.

Where possible the designated outline should exclude common items such as pavements, drains, etc., but where this cannot be done a schedule of work to be excluded from the lump sum must be provided.

The Contractor may find that his lump sum is inadequate when the detailed, post-Contract design work is carried out. He must judge how much detailed design to carry out in a tendering situation so as to avoid this with no certainty that he will be awarded the Contract.

The lump sum allows for consequential changes outside the DO. These may be difficult to assess.

The schedule of quantities will often not reflect the true rates for the work as they must total to the lump sum which the detailed design may show to be inaccurate.

The schedule is for the Engineer's acceptance (para. 16 of Preamble) but there is no indication of the position if the Engineer is dissatisfied with the rates or on what basis acceptance can be withheld.

These new procedures are being introduced for a limited range of relatively simple structures where proprietary products appear to be competitive. These are:

precast concrete box culverts up to 8 m span
corrugated steel structures (0.9 to 8 m span)*

drains (exceeding 0.9 m diameter)

reinforced earth structures*
crib walling
anchored earth structures

environmental barriers
footbridges
small span underbridges (up to 8 m span)

* pre-contract type approval required

In the case of structural elements, similar procedures have been adopted to avoid the 'barriers to trade' difficulty but without the complications of an O/AIP and designated outlines. Thus stuctural elements or other features should not be specified if they are based on proprietary systems and instead such elements must be designed by the Contractor (or manufacturer) and listed in Appendix 1/11. Examples are:

combined drainage and kerb systems
ground anchorages for anchored structures
piles
bridge bearings
bridge expansion joints
parapets

The Engineer should be aware that under the MCD amendment to ICE Clause 8 he becomes responsible for the design of these structures and structural elements on completion of his check (see Box 5).

The above changes have largely been brought about as a response to the Treaty of Rome but other changes have been made in the 100 series as a result of problems on previous contracts. The most substantial of these changes has been to the sampling and testing aspects of **Clause 105** and the quality assurance aspects of **Clause 106**.

The Sixth Edition introduced as an option the testing of materials or of the finished work by the Contractor whilst at the same time the Third Edition of the MMHW no longer itemised testing in the Bill of Quantities.

> 8B(3) The Engineer shall accept responsibility for the Contractor's design after the check procedures and shall notify the Contractor of the date of acceptance.

Box 5 Amended ICE Clause 8

Thus the payment for any tests carried out by the Contractor rested on Clause 36(3) of the Conditions of Contract which states 'The cost of making any test shall be borne by the Contractor if such test is clearly intended by or provided for in the Contract . . .'.

Unfortunately, due to imprecise wording of the SHW clauses it was often not clear as to what tests the Contractor had to carry out and as a result there were often disputes over payment. The Contractor is also required to provide samples for the Engineer to test and, as the wording of the Conditions of Contract Clause 36(2) is the same as that for tests, similar payment problems can arise.

The Seventh Edition has addressed this problem by providing more guidance on how Contractor testing can be used and by introducing two new appendices (1/5 and 1/6) which should list all the tests the Contractor is required to carry out and the samples he is to provide for the Engineer's testing. In addition an amendment to ICE Clause 36(3) is introduced through the Model Contract Documents (see Box 13) and a table of suggested tests is provided in Notes for Guidance Table 1/1.

Whilst properly completed Appendices 1/5 and 1/6 should simplify the pricing of these tests and remove any uncertainty regarding payment, difficulties are likely to arise if the frequency or type of test or sample is varied during construction. In this situation the Engineer would need to issue an ordered variation, but there are no Bill of Quantities rates to assist in valuing the variation.

Traditionally the Engineer on trunk road contracts has undertaken the testing, usually through a firm of testing sub-consultants or a County Highways Laboratory. Whilst the DTp. do not anticipate any major change in responsibility for testing (i.e. that it will generally remain with the Engineer) the Seventh Edition allows the designer to specify which tests the Contractor must carry out and these are selected from NG Table 1/1.

The Contractor, however, has an obligation under Clause 36(1) of ICE Conditions of Contract which states 'All materials and workmanship shall be of the respective kinds described in the Contract . . .'. If, in order to ensure that the materials or workmanship are as specified, the Contractor carries out tests he is merely complying with this requirement. Thus some Contractor testing is already provided for in the Conditions of Contract but without any specific details regarding frequency or number.

There would seem to be an area of potential ambiguity in the situation where the Engineer is responsible, say, for earthworks classification tests, as the obligation of the Contractor under Clause 36(1) seems diminished with possible problems in the event of mis-classification.

In the Sixth Edition, the DTp. introduced the policy of using NAMAS (NAtional Measurement Accredition Service — see Box 14) accredited laboratories, as it was considered there was sufficient capacity to make this a reasonable requirement. The DTp. are now turning their attention towards NAMAS accredited site laboratories for testing and sampling (at least on the larger trunk road contracts). A transition period has been allowed so that testing sub-consultants can gain accreditation and NAMAS accreditation can become a pre-qualification for testing laboratories on major schemes.

To ensure common standards for testing where Contractor testing is used there is a similar requirement that NAMAS accreditation should apply to all tests so specified in Appendix 1/5. (NG Table 1/1 indicates which tests should be so accredited.)

In the area of Quality Assurance the Seventh Edition has extended the concept introduced in the Sixth Edition by taking account of new schemes being applied by the industry. The DTp.'s policy has been to encourage QA and it has adopted QA schemes to BS 5750 where these have covered significant areas of trunk road work. The emphasis in the SHW is gradually changing towards a greater reliance on QA schemes to promote and encourage consistency of standards of work and materials.

QA involves an independent body, usually a third party, which provides confidence that specified standards of production are being met (see also Box 11). It does not replace the contractual obligation that work, goods and material must be as specified in the contract documents.

In the Sixth Edition all BS 5750 schemes were included in Appendix A, but in the Seventh Edition the Quality Management Schemes and Product Certification Schemes have been separated into different appendices (see notes on Clause 104 and Box 11).

Two new clauses are now included to improve the health and safety aspects of the SHW — **Clause 123** which governs the use of nuclear gauges on site and **Clause 124** which ensures that adequate safeguards are included in the Contract to protect the health of members of the public as well as that of site staff. The latter has particular relevance to the use of paints and the application of bridge deck waterproofing and silane.

Clause 117 — *Traffic Safety and Management* has been considerably updated and takes account of the new Chapter 8 of the Traffic Signs Manual and other operational and safety considerations. In particular it now allows for the design, construction, maintenance and removal of central reserve crossovers and includes the option of the Contractor to undertake maintenance functions on lengths of highway where the DTp. are the highway authority and there is a heavy flow of traffic.

There are also new requirements for site staff to wear jackets to BS 6629 (Class A to Appendix G) and provision is made for the appointment of a Traffic Safety and Control Officer to contol the operation of traffic safety and breakdown services.

In the following notes on individual clauses, the use of italics in headings, text and clause number denotes areas of change.

101 • Temporary Accommodation *and Equipment* for the Engineer

This clause has been considerably reworded with an improvement in clarity but without any major change.

Unless stated otherwise in Appendix 1/1 accommodation is required from four weeks after commencement to the issue of the Maintenance Certificate. MMHW separates 'before Completion' and 'after Completion' (see Box 6).

If accommodation is required during the first four weeks, this should be specified in Appendix 1/1 as temporary initial accommodation (see Box 6 for measurement aspects) or, alternatively, the main accommodation can be required from commencement if expressly stated. Any requirement for subsidiary or off-site accommodation should also be stipulated.

The confusing advice regarding provision of offices before the main accommodation is available or after completion of the works has now been omitted.

The compiler should list the furniture and equipment, the standard of artificial lighting and the minimum room temperature during *stated hours*. Equipment and fittings need not necessarily be new. *Testing equipment listed should only include that required for use by the Engineer for compliance testing (in particular the tests specified in Appendix 1/6) and test equipment must be properly calibrated by the Contractor. The Engineer should ensure that this is properly done as failure to do so might invalidate NAMAS accreditation (see Introduction to this series and Box 14).*

If the Engineer's accommodation is erected on part of the site, or on land with a common boundary to the site, planning permission is deemed to have been granted for the duration of the construction operations. ICE Conditions of Contract Clause 26(2)(c) warrants that the Employer has obtained Planning Permission for the Permanent Works or any Temporary Works specified by the Engineer. The Engineer will need to take care that planning permission for the temporary accommodation specified by him has the necessary approval.

Temporary Accommodation — Itemisation

3 Separate items shall be provided for temporary accommodation in accordance with Chapter II paragraphs 3 and 4 and the following:

Group	Feature	
I	1	Erection.
	2	Servicing.
	3	Dismantling.
II	1	Principal offices for the Engineer.
	2	Principal laboratories for the Engineer.
	3	Portable offices for the Engineer.
	4	Portable laboratories for the Engineer.
	5	Offices and messes for the Contractor.
	6	Stores and workshops for the Contractor.
III	1	Provided by the Employer.
IV	1	At the place of fabrication or manufacture.
V	1	Until completion of the Works.
	2	After completion of the Works.

Note: Group IV and Group V features shall be applied only to items of temporary accommodation for the Engineer.

Erection of temporary accommodation — Item coverage

— In the case of accommodation for the Engineer:

(a) initial accommodation and equipment, maintenance, servicing and removing;

Note that initial accommodation is not separately measured

Box 6 MMHW — Temporary Accommodation

102 • Vehicles for the Engineer

The period for which vehicles are required should be stated in Appendix 1/2 together with vehicle type. The description should not rely on proprietary names but if this is unavoidable the words 'or equivalent' should be added (see Box 7 for MMHW rules).

The Notes for Guidance are considerably changed in this respect. 'Land Rover type four-wheel drive or equivalent' has disappeared and a general vehicle specification provided. Additional requirements are added regarding 'free from markings' and the 'Light Van' need no longer be front-wheel drive with a four-cylinder engine.

New vehicles should only be specified where the 'nature and contract period make it essential' *and the Overseeing Department have agreed this with the Engineer*. The vehicles are to be delivered and maintained in good roadworthy condition and if out of service for more than 24 hours a suitable replacement must be provided.

The Contractor must have comprehensive insurance for the vehicles for any qualified driver authorised by the Engineer and for authorised passengers.

8 The measurement of vehicles for the Engineer shall be each week or part thereof during which a vehicle is provided.

Box 7 MMHW — Vehicle Weeks

103 • *Communication System for the Engineer*

This clause has been reworded to allow communication systems other than radio.

The scale of provision is approximately one set per kilometre. If the equipment is required earlier than four weeks after commencement this should be stated in Appendix 1/3. *The equipment should be removed at the end of the period stated in Appendix 1/3 rather than on completion of the Contract*. The frequency should be obtained by the Engineer from the Employer *if radio communication is used.*

The acceptance of a PEQ as equivalent in terms of its safety, suitability and fitness for purpose should be assessed under the following:

(a) Standards — a PEQ meeting the requirements of a National standard of a member state for use in similar circumstances should normally be accepted unless it has markedly lower levels of safety, suitability and fitness for purpose. The factors underlying the purpose of the specification and their criticality should be evaluated so far as is practical.

(b) Quality Assurance Schemes — when ascertaining whether or not a proposed quality management or product certification scheme of another member state is equivalent, the Engineer should consider those factors forming the basis of the acceptability of the listed scheme. The Engineer should check that:

(i) the certification being offered is from an independent body;
(ii) the certification body has received accreditation for that scheme (e.g.NACCB recommends certification bodies for accreditation to DTI)
(iii) in the case of product certification schemes that test houses are independent, suitable test equipment is being used, the correct tests are being undertaken and that the sampling frequency gives a similar degree of confidence (i.e. that product testing is fully equivalent).

It should also be noted that any testing of a PEQ which has been accepted should be in accordance with the standard or technical specification to which the PEQ conforms, providing that the information from such tests is comparable to that required in the contract for assessing suitability

Box 8 Acceptance of a PEQ — DTp. advice

104 • *Standards, Quality Assurance Schemes. Agrément Certificates and other Approvals*

This clause and its associated Notes for Guidance has been completely rewritten and considerably extended compared with the Sixth Edition and introduces the principle of reciprocity into Overseeing Department contracts.

It sets out the requirements which need to be met for work, goods and materials which are to be used in the works. The ideal would be to specify the work in terms of performance and to assess compliance against the end result but this is not possible in many cases. The specification therefore relies on specific standards, QA schemes, Agrément certificates and other approvals. This clause sets out the requirements for these but most importantly sets out the principle of equivalence.

***Equivalence** entails the acceptance of 'equivalent' work, goods and materials meeting an equivalent standard, QA scheme, Agrément certificate or approval of any member state of the European Communities. This does not mean that they must be identical but that they offer similar levels of safety, suitability and fitness for purpose.*

*In general, where particular standards, quality management, product certification schemes and Agrément certification are required the Contractor can comply if the goods or materials satisfy the similar requirements of a member state of the EC. He does so by proposing that goods or materials are equivalent — this **PEQ** (proposed equivalent goods and materials) must then be assessed by the Engineer who will decide if it is equivalent (see Box 8). The Engineer would be advised to ensure that the Contractor's proposals are a PEQ under the equivalence rules rather than an alternative to the Engineer's design. In the latter case the Instructions for Tendering require the tenderer to bear the cost of the independent check.*

The Contractor is required to provide all the relevant information (such as specifications, certificates, test data and inspection reports) to the Engineer so that he can ascertain whether the proposal is equivalent (see the 'Provision of Information' requirements of this clause).

The Engineer, in considering whether the proposed standards, etc., are equivalent should consider the underlying purpose of the specified requirement and not deny equivalence because of a detail which does not

Equivalent Products and Materials

14 Where the Contractor offers an equivalent product or material in place of the one identified or specified, which is accepted for incorporation into the Works by the Engineer, then the rates and prices in the Bill of Quantities shall be deemed to include for all the obligations and costs associated with the incorporation of the equivalent into the Works, including design, provision of data and drawings, certificates, awaiting approvals, resubmissions and modifications and amendments to the Works.

Unless specifically stated to the contrary in the Contract the measurement of the Works affected by the incorporation of the equivalent products and materials shall be based on the Tender documents and not on the Works as amended and completed to incorporate the equivalent products and materials.

Box 9 MMHW — Preambles to Bills of Quantities

essentially affect this. Factors such as consequential effects on design and cost should also be considered and the Preambles to Bill of Quantities now covers this (see Box 9).

Advice Note SA 2/92 offers advice on assessing equivalence in general terms with procedures to be followed so that a consistent and logical approach is achieved which avoids unnecessary duplication. The DTp. can be asked for assistance in the Engineer's assessment of equivalence.

In places, the specification requires work, goods and materials to have received Statutory or Departmental type approval/registration before they can be incorporated into the works (see Box 10). Whilst similar equivalence rules apply, it is anticipated that most PEQ's submitted in this category will be outside of a contract and will be requested by a manufacturer, supplier or agent (Box 10). Whilst this does not preclude a contractor putting forward a PEQ requiring such type approval/registration, the procedures for obtaining this can take a considerable time and the contractor will need to be aware of this. SD 1/92 advises the Engineer to inform the Contractor of the timescale.

Statutory type approval and statutory authorisation apply to traffic signs and the Secretary of State grants the approval for the Overseeing Department. Appendix D lists signs requiring Statutory type approval — e.g. variable message signs, reflecting road studs and traffic signal and control equipment.

Statutory authorisation is required if signs are to be erected which are of a character or are used in circumstances not prescribed by the Traffic Signs Regulations and General Directions 1981 and the Pelican Crossing Regulations and General Directions 1987.

Departmental type approval is granted by the Overseeing Department and Appendix E lists work, goods and materials subject to such requirements — e.g. safety fences, bridge expansion joints and materials for corrugated buried structures. The Contractor is required to supply a copy of the certificate to the Engineer.

Annex A of Departmental Standard SD 1/92 lists the work goods and materials which require Statutory or Departmental type approval/registration and gives an indication of the timescale needed for approval/registration to be obtained.

Advice Note SA 1/92 lists those products that have obtained Statutory type approval or Departmental type approval/registration together with any restrictions or limitations on their use. It will be updated on a regular basis and enquiries may be made of the Overseeing Department to check that any approval is current or whether other products have been added subsequent to the latest publication of the advice note.

Box 10 Information available on type approval, etc.

It should be noted that there is some confused wording in the SHW regarding the procedure for assessing equivalence in the case of departmental type approval. Clause 104.13 states 'When the Contractor proposes to use a different . . . Departmental type approval/registration from that specified the Contractor shall provide all the information . . . to enable the ***Engineer to ascertain*** *whether or not the proposal is equivalent . . .'. However, SHW Clause 104.10 states 'Where there is a requirement . . . to have Departmental type approval/registration this will be* ***granted by the Overseeing Department*** *where . . . have an equivalent approval of the national highway authority of any member state . . .'.*

Equivalence rules apply to standards only where the standard specified is other than a Harmonised European Standard, or technically equivalent to one. Compliance can be through a relevant standard, code of practice, recognised international standard, acknowledged technical standard or written description of traditional procedures.

Quality Assurance (QA) *schemes to BS 5750 Part 2 were first introduced in the Sixth Edition (e.g. for reinforcing steels, bridge parapets and safety fencing) and the Seventh Edition has added more, taking account of new schemes being applied by the industry (see Box 11 for an explanation of QA).*

In the Sixth Edition all BS 5750 schemes were included in Appendix A, but in the Seventh Edition the opportunity has been taken to separate into different appendices Quality Management Schemes and Product Certification Schemes. Thus Appendix A now lists only Quality Management Schemes (e.g. Portland cement, fencing and safety fencing, timber preservation, parapets, lighting columns, road markings and traffic signals) and Appendix B lists Product Certification Schemes (e.g. Kitemark — drainage products, Safety Mark — lighting equipment, Other Marked Schemes — electric cables and reinforcing steel, and Non-Marked Schemes — ready mixed concrete).

SHW Clause 104.3 requires the Contractor to use only work, goods or materials in the Works if such work, etc. is the subject of an accepted quality assurance scheme listed in Appendix A or B (see Box 12 for MMHW aspects). The Contractor is then required to submit to the Engineer a copy of the certificate of conformity which the Engineer should check and retain as evidence of the operation of the schemes (NG 104.4).

A quality management scheme does not guarantee the quality of the product as the schemes are solely based on management procedures. With

There are two types of Quality Assurance used in the SHW:

Quality Management schemes
Product Conformity certification

Additionally, there are **Departmental Type Approval/Registration** procedures some of which are statutory.

Quality Management schemes are those which comply with the BS 5750 standards, but do not involve a product being independently tested, only a presumption that if work is properly managed then clients' needs are being met (Appendix A).

Product certification involves a testing of the product. Examples are the Kitemark schemes operated by British Standards Institution Quality Assurance (BSI QA), where products are being certified as conforming to a product standard (Appendix B).

In the case of innovative products where there are no established standards approval is given on the basis that the manufacturing process and use of the product is being monitored. Examples are the British Board of Agrément (BBA) Roads and Bridge Certificates (Appendix C).

Equivalence rules apply to all of the above.

Box 11 Quality Assurance

Preambles to Bill of Quantities General Directions — 2

(xi) Complying with Quality Assurance schemes and providing certificates of conformity.

Box 12 MMHW — Quality Assurance

product certification schemes the materials will have been subject to independent testing which may allow the Engineer to reduce or eliminate inspection and testing (see NG 104.5).

British Board of Agrément Road and Bridges Certificates. *SHW Clause 104.5 requires the Contractor to use only work, goods or materials certificated by BBA where such work, etc. is required to have such certification and he must submit a copy of the certificate to the Engineer.*

Appendix C lists those types of work, goods and materials subject to such requirements. These are proprietary products which are required to have a BBA Roads and Bridges Certificate (e.g. permanent shuttering for road gullies, fin drains, waterproofing for bridge decks, protective coatings and paved inverts for corrugated buried structures and reinforced earth systems). (See also Box 11).

BBA is a member of the European Organisation of Technical Approval (EOTA). It is therefore possible to check whether a certificate issued to a manufacturer by another EOTA member offers equivalent guarantees of safety, suitability and fitness for purpose.

Statutory type approval *is a requirement of the Traffic Signs Regulations and General Directions 1981 and the Pelican Pedestrian Crossing Regulations and General Directions 1987 (see also Box 5).*

Statutory authorisation *is required if signs are to be erected which are of a character or are used in circumstances not prescribed by that legislation.*

Traffic signs requiring type approval are listed in Appendix D and the Contractor must provide the Engineer with written evidence that approval has been given. The Secretary of State for the Overseeing Department is responsible for such approval.

Goods and materials requiring ***Departmental Type Approval/ Registration*** *are listed in Appendix E and the Contractor must submit a copy of the certificate to the Engineer. Equivalence rules relate to the 'national highway authority' of any member state and the Engineer must forward equivalence information, provided by the Contractor, to the Overseeing Department in time to allow for approval (see note above regarding the confused wording of SHW Clause 104.10 and 104.13).*

Provision of Information. *The requirements of sub-clauses 11 and 12 are broadly similar to those in clause 107 and Appendix 1/7 of the Sixth Edition.*

At least four weeks prior to their use the Contractor submits to the Engineer two copies of information in respect of goods, etc. which he proposes to use — where necessary with an English translation.

At least four weeks prior to commencing the work the Contractor submits three copies of working and fabrication drawings (where required in Appendix 1/4). These could include steelwork, parapets, waterproofing details, diaphragm wall details, traffic signs, lighting, bearings, precast concrete, joints, environmental barriers, corrugated steel buried structures, combined drainage and kerb systems. These are submitted for the Engineer's approval following which the Contractor provides transparencies of the approved drawings.

Due to the greater emphasis on the Contractor's design in respect of proprietary materials there is likely to be much greater use of this aspect of the specification.

105 • *Goods, Materials, Sampling and Testing*

This clause and its associated Notes for Guidance clause is completely rewritten and two new Sample Appendices have been introduced (see also notes in the Introduction to this series).

Goods and Materials. *The Contractor must submit a list of his proposed suppliers to the Engineer. Nothing is said regarding the need for the*

Clause 36

Sub-clause (3) is deleted and replaced by the following Sub-clause

Tests

(3) Each test of materials or workmanship which is specified in the Contract as to be carried out by the Contractor shall be carried out by him at his own cost provided that the test in question is particularised in the Specification in sufficient detail to enable the Contractor to have priced or allowed for the same in his Tender. The cost of any test carried out which is:

(a) not so particularised in the Specification; or

(b) specified in the Contract and carried out by the Engineer; or

(c) not specified in the Contract;

shall be borne by the Contractor if the test shows the workmanship or materials not to be in accordance with the provisions of the Contract or the Engineer's instructions, but otherwise by the Employer.

Box 13 MCD — Amendments to Conditions of Contract Clause 36

Engineer's approval but Sub-clause (1) goes on to say that no change shall be made in the list and the Contractor's proposals without the Engineer's prior approval.

Samples — which should be identified in Appendix 1/6 — should be retained by the Engineer until the completion of the works (NG 105.1). It is unclear whether this means substantial completion or the issue of the maintenance certificate.

Sampling and Testing. *In order to avoid the difficulties over payment which arose in the Sixth Edition in respect of the wording of ICE Clause 36, where the cost of any test is borne by the Contractor if such test 'is clearly intended by or provided for in the Contract', the MCD have modified the Conditions of Contract clause (see Box 13).*

Appendix 1/5 is intended to identify clearly the tests and enable the Contractor to price for carrying them out. The designer can elect for the Contractor to carry out most of the testing with the Engineer then doing only audit testing, although the DTp. do not anticipate any major change from the traditional approach of the Engineer being responsible for most testing. The Contractor must be required to carry out those tests indicated as such in Table NG 1/1 otherwise conflict with SHW clauses may arise (see also the Introduction to this series).

The Seventh Edition has extended the NAMAS accreditation of testing introduced in the Sixth Edition (see also Introduction to this series and Box 14) and in order to ensure common standards Appendix 1/5 will identify where test certificates are required and where tests must be carried out by NAMAS accredited laboratories. The Contractor may propose to have tests carried out on his behalf by a testing laboratory, manufacturer or supplier (NG 105.2).

Notes for Guidance Table 1/1 give advice for the compiler in completing Appendix 1/5 but other tests, e.g. those arising from contract-specific clauses should be included as appropriate. The list of tests is not exhaustive, nor is it intended to be used without thought by the designer — over-specification of requirements for tests is wasteful of resources and there may be cases where the suggested testing rates need to be increased where the Contractor is likely to be using marginal materials.

When completing Appendices 1/5 and 1/6 the compiler should consider the situation where a Contractor may be responsible for the design of part of the permanent works or may have an option of using different materials, e.g. different pavements or pipes for drainage. In such a case he should ensure that the tests and samples listed cover all of the possible options allowed.

Quality Assurance of testing is organised through NAMAS (NAtional Measurement Accreditation Service) which is an agency of the National Physical Laboratory.

It accredits laboratories for testing and for sampling and it has been the DTp.'s policy since 1989 to use NAMAS accredited laboratories wherever practicable.

Recently the Engineer's laboratories on larger contracts (over £5 million) have begun to be NAMAS accredited.

Box 14 NAMAS

Preambles to Bill of Quantities

General Directions — 2

(vii) General obligations, liabilities and risks involved in the execution of the Works set forth or reasonably implied in the documents on which the tender is based

(x) Attendance and transport for sampling and testing carried out by the Engineer, supplying results of tests carried out by the Contractor and providing test certificates.

Box 15 MMHW — General Directions applicable to testing

Where NG Table 1/1 refers to a SHW clause for the frequency of the test this frequency must be repeated in Appendix 1/5.

In the case of samples, the amended Clause 36 (see Box 13) does not change the applicable wording which appears in ICE Clause 36 (2): 'All samples shall be supplied by the Contractor at his own cost if the supply thereof is clearly intended by or provided for in the Contract but if not then at the cost of the Employer.' The Contractor must supply the samples as well as carrying out the tests described in Appendix 1/5 but where the Engineer requires samples to be provided for his own testing, these must be identified in Appendix 1/6. This should include details of the frequency of sampling and the delivery location. The Contractor must provide these samples in 'sufficient time for them to be tested and approved by the Engineer taking into account the programme for the Works'.

The Contractor is required by SHW Clause 105.2 to obtain test certificates (as the purchaser) provided for in a British Standard or other standard where this is stated in Appendix 1/5. They must be provided at least four weeks before being incorporated into the Works.

If during the construction of the Works it is neccessary for the Engineer to increase the rate of testing then, as this cannot have been allowed for by the Contractor in his prices, he will need to issue an ordered variation, the valuing of which is likely to give rise to difficulties as there will be no comparable items in the Bill of Quantities.

Where materials are accepted on the basis of an equivalent standard, etc. as allowed for in clause 104, the testing and sampling applicable to that standard 'is accepted and shall be substituted for those specified in Appendix 1/5 and 1/6 respectively' (SHW Clause 105.4).

Where testing is carried out in another member state of the EC, the tests should be carried out by an 'appropriate organisation offering suitable and satisfactory evidence of technical and professional competence and independence'.

106 • *Design of Permanent Works by the Contractor*

This is a completely new clause with the requirements of the Sixth Edition Clause 106 Alternative Specified Materials being incorporated into clause 105.1.

The need for this clause has arisen largely from the need to avoid 'barriers to trade' as interpreted from Article 30 of the Treaty of Rome by naming or otherwise favouring particular proprietary materials (see also the Introduction to this series). In addition, the DTp. aim to encourage the use of proprietary materials where they may offer value for money and this clause is designed to encourage that aspect also.

In the past there has been some confusion as to how such structures or elements should be incorporated into the Contract and there has been legitimate cause for complaint with some cases going to the European Commission. The DTp. have therefore developed a new approach acceptable to the EC and the procedure for this is set out in new Departmental Standard SD 4/92 and is summarised in Box 3.

Thus where any structure is substantially based on proprietary materials or systems it must not be fully detailed in the Contract. It should be listed in Appendix 1/10(A) and the Contractor is then required to carry out the detailed design in compliance with the design requirements referred to in the appendix. Where there is a choice of designs then the requirements are to be listed in Appendix 1/10(B).

All proprietary manufactured structures still require specific technical approval to BD 2/89 Technical Approval for Highway Structures but the new procedures recognise and address the EC philosophy in avoiding discrimination.

Essentially the Contractor now has to comply only with an outline

Essential Requirements of an 'Outline Approval in Principle' — (O/AIP)

1 Location
2 Operational dimensions/levels
3 Highway loading requirements
4 Other loading requirements
5 Relevant Departmental Standards, British Standards, Codes of Practice, etc.
6 General arrangement drawing including the Designated Outline
7 Ground investigation data
8 Appearance of structure
9 Environmental factors
10 Constraints/external control during construction
11 Operational or user requirements
12 Special maintenance requirements
13 Other essential requirements

Box 16 Essential requirements of an O/AIP

specification which should give scope for the choice of different proprietary structures whilst still meeting the Engineer's design criteria. He tenders on a lump sum basis and completes the detailed design after award of the Contract.

The Outline Approval in Principle form (O/AIP) requires technical approval by the TAA and the essential requirements are shown in Box 16. In compiling the O/AIP the Engineer should leave as wide a choice as possible and in general terms 'the least said the better' about the actual structure, as long as the design requirements are clearly specified and it is clear where it goes and how it links up to the other parts of the Works.

Appearance and environmental factors must be descibed only in performance terms (otherwise this could constitute a 'barrier to trade') and the fullest possible ground investigation data must be made available together with any constraints with regard to timing, traffic, etc.

The Contractor is required to complete the Outline Approval in Principle form (which should be included in the design specification) and forward it to the Engineer for acceptance (SHW 106.2). This wording is rather misleading as the Technical Approval Authority, not the Engineer, accept it and the Notes for Guidance advise the Engineer to pass it to 'the Technical Approval Authority for acceptance'.

The Engineer will be expected to check the Contractor's design prior to passing it to the TAA. After technical acceptance, the Enginner 'takes over' the structure. The Model Contract Documents have modified ICE Conditions of Contract Clause 8 substantially to cover these changes in the specification (see Box 17).

Such structures would include environmental barriers, crib walling, precast concrete box culverts (up to 8 m span), corrugated steel buried structures (0.9 to 8 m span), reinforced earth structures, anchored earth structures, footbridges, small span underbridges (up to 8 m span) and drains (exceeding 0.9 m diameter).

For the same reasons the Engineer should not specify structural elements or other features if they are based on proprietary materials. These are to be listed in Appendix 1/11 and the Contractor is then required to design them in compliance with the design specifications referred to in the appendix and submit his proposals to the Engineer for approval. The Contractor can propose an element designed by the manufacturer. Such structural elements would include combined drainage and kerb systems, ground anchorages for anchored structures, piles, bridge bearings, bridge expansion joints and parapets. The appendix should also include any non-proprietary structural elements required to be designed by Contractor, e.g. foundations to lighting columns.

Contractor's Design Responsibilities

8B (1) The Contractor shall design parts of the Permanent Works as required in accordance with the provisions of the Specification and submit drawings and specifications of his design to the Engineer. The Contractor may submit a design prepared on his behalf by a sub-contractor or professional designer or propose a design prepared by a manufacturer.

(2) The Engineer shall examine and check the Contractor's design or proposal and inform the Contractor in writing within a reasonable period after receipt of full particulars either:

(a) that the design or proposal has the approval of the Engineer, or
(b) in what respects in the opinion of the Engineer the design or proposal fails to meet the requirements of the Specification.

In the latter event the Contractor shall take such steps or make such changes to the design or proposal as may be necessary to meet the Engineer's requirements and to obtain his approval.

(3) the Engineer shall accept responsibility for the Contractor's design after the check procedures and shall notify the Contractor of the date of acceptance.

(4) Unless otherwise agreed by the Engineer parts of the Permanent Works designed by the Contractor shall not be incorporated into the Works until the Engineer has informed the Contractor in writing that the design or proposal has the approval of the Engineer.

(5) The cost of the Engineer's examination and check of the Contractor's design or proposal shall be borne by the Employer.

Box 17 MCD Amended ICE Conditions of Contract Clause 8

In the case of structures designed by the Contractor it is obviously impractical to bill these in the normal way as details are unknown. To overcome this the drawings should show a Designated Outline for each structure that the Contractor is required to design and his lump sum price will relate to this (see also the Introduction to this series) .

This Designated Outline (DO) specifies the boundary to the lump sum and must be large enough to accommodate any likely options. It should not be larger than necessary and should exclude common items of construction, e.g. a culvert would exclude the surfacing but a bridge with different surfacing from the roadworks would include it.

Unaffected items of work can be excluded, e.g. drainage, kerbing, safety fencing etc. but the extent of the work excluded must be clearly shown and scheduled as it will affect the lump sum. Earthworks within the DO should not be included by the Engineer in the earthworks schedule as they are not measured in the 600 series.

A new section of the MMHW (Series 2500) sets out the measurement rules and appropriate changes have also been made to the 'Preparation of Bill of Quantities' and to the 'Preambles to Bills of Quantities' (see Boxes 4 and 18).

To cater for the valuation of variations within the DO the Contractor is required to provide a schedule of rates which must total to the lump sum inserted in the tender (see Box 4).

The Contractor is also required to propose lighting columns and brackets which have been designed by the manufacturer in accordance with Standard BD 2/79 Part IV and the 1300 series. The amended ICE Clause 8 would seem to require the Engineer to accept responsibility for the Contractor's design although the procedure under the Sixth Edition does not require this.

107 • *Site Extent and Limitations on Use*

This is a completely new clause with the requirement of the Sixth Edition clause 107 'Working and Fabrication Drawings and Manufacturers

Preparation of Bill of Quantities

8 Structures Designed by the Contractor

Where the Contract provides only for a structure designed by the Contractor to be constructed a Bill of Quantities comprising a single item for all the work within the Designated Outline (with the exception of those works scheduled as not to be included) is to be provided in accordance with Series 2500. Those works scheduled as not to be included in this single item shall be included by the Engineer in other Bills compiled in accordance with the appropriate Series.

Preambles to Bill of Quantities

Permanent Works Designed by the Contractor

15 Where the Contract requires part(s) of the Permanent Works to be designed by the Contractor, the rates and prices in the Bill of Quantities shall include for all the obligations and costs associated with the incorporation of the Contractor's design into the Works, including design, provision of data and drawings, certificates, awaiting approvals, resubmissions and modifications and amendments to the Works.

Box 18 MMHW — Structures Designed by the Contractor

Instructions' and its associated Appendix 1/7 now being incorporated into clause 104.14 and 104.15 together with Appendix 1/4.

The clause requires the extent of the site and any limitations on its use to be set out in Appendix 1/7. A new paragraph (12) has been added to the Preambles to Bill of Quantities to cover this requirement. This will effectively be an extension of the definition of site contained in the ICE Conditions of Contract Clause 1(1)(n): 'Site means the lands and other places on under in or through which the Works are to be executed and any other lands or places provided by the Employer for the purposes of the Contract'.

The extent of the site can be shown on the drawings with reference to them in Appendix 1/7 or where appropriate it can be described in the appendix. The extent should include those areas of highway necessary for the Contractor's temporary signs, cones and road markings which may be outside the area of the permanent works.

108 • Operatives for the Engineer

There are no major changes in this clause.

Appendix 1/8 specifies details, including the period operatives are required. The clause also stipulates that survey and laboratory operatives must be experienced in such work. Under MMHW, continuous periods of less than four hours in any one day do not count in measurement. The item coverage includes periods of less than 4 hours but this must be read in conjunction with the basic principles of the MMHW.

109 • *Control of Noise and Vibration*

The Contractor must comply with any specific requirements of Appendix 1/9 and with BS 5228 *Parts 1, 2 and 4 for practical measures. Additional requirements are included in respect of control of vibration set out in Appendices 1/9, 2/4, 6/3, 6/13 and clause 607 'Explosives and Blasting for Excavation'.*

The Sample Appendix 1/9 is virtually unchanged in respect of the noise requirements and retains the vagueness and ambiguities of the Sixth Edition appendix. The compiler should agree measures with the Local Authority and insert these in Appendix 1/9 but it is for the Contractor to decide if he wishes to seek Local Authority consent.

The measures which have been informally agreed between the Engineer and the Local Authority should be shown in Sub-clauses 2, 3, 4, and the table, but this leaves the exact procedure unclear — does 'consent' in Sub-clauses 2 and 3 refer to Engineer or Local Authority? (Sub-clause 4 seems to refer to permission from the Engineer after consultation with Local Authority.)

The restrictions on working hours could considerably affect the Contractor's proposed method of carrying out the works and affect the Engineer's approval of the programme but it is not clear from the Sample Appendix whether these are requirements of the contract or recommendations.

The Notes for Guidance point out the powers of the Local Authority under the Control of Pollution Act 1974 particularly regarding plant and machinery, hours of work and level of noise.

A new SHW sub-clause states that compliance with the specific requirements of the Appendices does not confer 'immunity from relevant legal requirements'.

Drawings should show position of noise control stations and the schedule should show ambient noise levels. Note that Sub-clause 4 requires the ambient noise level from all sources not to exceed the appropriate level, etc. This could presumably include noise which is not the responsibility of the Contractor.

Any requirements for the control of vibration are to be shown in Appendix 1/9 with details of locations, limits of amplitude, instrumentation requirements and any arrangements by the Engineer to enable the Contractor to monitor vibration outside the site.

110 • Information Boards

Information boards should be erected within the highway boundary *at the locations and to the specification set out in Appendix 1/21 cross-referenced to the drawings*. Planning permission is not required if the work is on a trunk road but the Local Planning Authority should be informed as a matter of courtesy.

The Engineer should check that safety fencing has been detailed at the location as appropriate (NG 110.1).

Information boards are to be kept clean and maintained *in a safe and legible condition and removed on completion of the Works. (It is not clear whether this can be interpreted as substantial completion).*

111 • Existing Ground Levels

Existing ground levels are to be described in Appendix 1/12 by references to drawings. The Contractor is to satisfy himself and dispute any levels he considers incorrect. Presumably he will dispute only those advantageous to him. Ground level should not be disturbed before the Engineer's decision on a dispute. If this delays work the Engineer may be responsible.

112 • Setting Out

Setting out information should be made available to tenderers with a copy to the Contractor when the contract is awarded (NG 112.3). The Sample Appendix itself merely states that the information will be supplied to the Contractor 'at the commencement of the Works'.

The Contractor must check markers within three weeks of commencement and supply a list of those he considers to be in error to the Engineer who may agree revised values. This could have payment implications for the Employer (Clause 17 ICE Conditions of Contract). *The Contractor must bring to the attention of the Engineer any missing markers and must also comply with any specific requrements of Appendix 1/12.*

The Notes for Guidance advise the Engineer to check the setting out as the work proceeds. This could have implications for the Engineer if he failed to do so and the Employer suffered delay or incurred cost.

The Contractor is also responsible for any setting out necessitated by the diversion of statutory undertakers or other services.

The references to BIPS3 and DGM are no longer in the Sample Appendix.

113 • Programme of *Works*

The SHW clause is reworded and has a new title but, as previously, it refers the Contractor to Appendix 1/13 for all the detailed requirements. The Notes for Guidance offer sensible advice with only minor changes to the previous wording. Unfortunately some of the requirements as set out in Sample Appendix 1/13 are not consistent with these Notes for Guidance and cause problems when considered with Clause 14 of ICE Conditions.

It is desirable that the information required by the Engineer should be specified in line with NG 113.2 but the suggested format may lead to difficulties. Whilst the first two paragraphs of the sample appendix are desirable additional requirements to those in Clause 14 of the ICE Conditions of Contract, the amount of detail required for Level 1 is likely to present the Engineer with difficulties in approving the programme. He is being asked to approve the Contractor's programme without any of the detail specified for Level 2, some of which would normally be regarded as desirable or essential for the proper consideration of the Contractor's intentions.

The position with regard to Level 2 is not covered by Clause 14 of the ICE Conditions of Contract and its contractual significance and the question of the Engineer's approval (or acceptance) is therefore unclear.

Some additional constraints are now included, but there will often be Employer's constraints on the Contractor in respect of his programme which have not been mentioned in the Sample Appendix.

114 • Monthly Statements

The wording of this Clause has been extended to refer to the requirements of the ICE Conditions of Contract. Appendix 1/14 sets out the detailed arrangements for the Contractor's monthly statements which supplement the requirements of Clause 60(1) of the ICE Conditions of Contract.

115 • Accommodation Works

Details of accommodation works should be shown in Appendix 1/15 indicating the periods for completion. *The misleading reference to 'within such periods of time as may be required by the Engineer' is now omitted from the SHW clause. It is now clear that any constraint in respect of completion should be stated in Appendix 1/15 and presumably also in Appendix 1/13 although this is not listed in Sample Appendix 1/13 as a constraint.*

The specification also requires the Contractor to give at least ten days' notice of the start of work. As accommodation works commonly include the boundary fencing this could have implications in respect of temporary fencing requirements.

The Notes for Guidance and the Sample Appendix are now clearer on the situation where accommodation works have not been finalised prior to inviting tenders.

116 • Privately and Publicly Owned Services or Supplies

This clause is considerably reworded with the ambiguous requirements of the Sixth Edition regarding the provision of satisfactory alternatives for privately owned services being omitted.

The responsibility of the Contractor regarding phasing of these works

> Preambles to Bill of Quantities
>
> General Directions
>
> 2(vi) The effect on the phasing of the Works or any element of the Works to the extent set forth or reasonably implied in the documents on which the tender is based.

Box 19 MMHW Preambles to B of Q

was previously only referred to in the Notes for Guidance although the Preamble to Bill of Quantities in the MMHW covered this in para. 1(vi). These requirements have now been moved into the SHW where they form part of the contract documents and the wording of the Preamble has been made less specific (see Box 19). This requirement is to be read in conjunction with the information on notice and time for completion required to be shown in the appendix by the Engineer.

The Contractor should take all measures required, etc. for the support and protection of services. *Additional wording qualifies this as being subject to the instructions or contrary directions of the Engineer, payment in this case depending on the relevant Conditions of Contract clause.* The Preamble to the Bill covers this but only to the extent that such work 'is set forth or reasonably implied in the documents on which the tender is based'.

Preliminary arrangements made by the Engineer with the Statutory Undertakers should be shown in Appendix 1/16 together with details of advance contracts, agreements or pre-ordered materials. Note that the Sample Appendix states that 'Compliance with the period of notice given in this Appendix does not relieve the Contractor of his obligations'. This needs to be read in conjunction with the ICE Conditions of Contract and the MMHW.

As this will affect the Contractor's programme they should also be shown as constraints in Appendix 1/13 as indicated in the Sample Appendix.

117 • Traffic Safety and Management

This extensive clause is considerably changed with an improvement in clarity and many additional requirements. It takes account of the new edition of Chapter 8 of the Traffic Signs Manual and additional safety considerations.

There is an important change in the MMHW in respect of 'contraflow' arrangements — the Third Edition MMHW seemed to require these to be itemised as temporary diversions of traffic although this was often overlooked, particularly when the Contractor was designing the arrangement. This caused difficulties over payment and the Fourth Edition MMHW has attempted to overcome this by requiring a separate item under Traffic Safety and Management when contraflow is to be adopted (see Boxes

> 21 The units of measurement shall be:
>
> (i) traffic safety and management . . . item.
> *(ii) taking measures for or construction, maintenance, removal of contraflow arrangements . . . item.*
>
> Measurement
>
> 22 Traffic safety and management shall be measured once only.
>
> 23 *Unless expressly stated otherwise in the Contract, taking measures for or construction, maintenance, removal of contraflow arrangements shall be measured once only to include for all contraflow arrangements specified by the Engineer in Appendix 1/17 and all contraflow arrangements proposed by the Contractor.*
>
> Itemisation
>
> 24 Separate items shall be provided for traffic safety and management in accordance with Chapter II paragraphs 3 and 4 and the following:

Group	Feature
I	1 Traffic safety and management.
	2 Taking measures for or construction, maintenance, removal of contraflow arrangements.

Box 20 MMHW — Contraflow Arrangements — introduced in the October 1991 Addendum

Taking Measures for or Construction, Maintenance, Removal of Contraflow Arrangements

26 Item coverage

(a) traffic safety and management (as this Series paragraph 25);
(b) temporary diversions for traffic (as this Series paragraphs 30, 31 and 32);
(c) crossovers;
(d) temporary removal and reinstatement;
(e) design of contraflow arrangements specified by the Engineer.

Box 21 MMHW Contraflow Item Coverage — introduced through October 1991 Addendum

20 and 21). The NG to MMHW recommends that the contraflow item is always included when traffic management is required thus allowing for the Contractor's proposals.

The Sixth Edition Sub-clauses (1) to (13) have been largely retained but with clearer wording and re- numbering. Thus, where previously reference was made to 'a trafficked highway' this now becomes 'a highway open to vehicles'. The Contractor's 'programme' now become 'proposals for traffic safety and management' and the Engineer can now require 'changes to his proposals as may be necessary, in the opinion of the Engineer, to meet the requirements of the Contract and to obtain consent'.

The Contractor must not commence any work which affects the public highway unless the traffic safety and management measures are operational and 'have been approved by the Engineer' (SHW 117.12). The Engineer's 'consent' is required in SHW 117.3.

Additional lettering x-heights are now allowed for vehicle sign boards and Appendix 1/17 should now show whether hazard warning lights are a permitted alternative to a roof mounted amber flashing light.

The requirements for 'personnel' to wear jackets has now been updated to comply with BS 6629 and clarified that this applies to 'the workforce and supervisory staff'.

New specification sub-clauses cover central reserve cross-overs which the Contractor may be required to design, construct and maintain as set out in Appendix 1/17. Full design requirements should be included in Appendix 1/17 and the Engineer should consult with the highway authority and list any maintenance requirements that the Contractor has to carry out (see Box 17 for changes to ICE Clause 8 in respect of Contractor's design. If the crossing is to remain it will presumably be part of the Permanent Works and therefore subject to this amended clause.)

In the Sixth Edition clause there was no specific reference to the Contractor's responsibility in respect of traffic orders and obtaining authority for temporary traffic signals, although the notice required by the Engineer to arrange these was stated in the Sample Appendix. Sub-clause 6 now corrects that omission and requires the Contractor to submit formally any statutory orders or notices to be published which are necessitated by his traffic safety and management proposals through the Engineer. Appendix 1/17 should stipulate the timescale for submission although unlike SHW Clause 118 this clause does not impose this restriction on the Contractor.

The Engineer should be informed of agreements which the Contractor has made with the highway authority for signs, lighting and central reserve cross-overs.

Specific provision is now made in the SHW for the possibility of the Contractor being responsible for maintenance functions (see Box 22 for changed item coverage). This would normally be where the Overseeing Department are the highway authority and the scheme involves the reconstruction or maintenance of a road carrying a heavy flow of vehicles.

Traffic Safety and Management 25 — Item Coverage Changes

(c) *initiating or continuing consultation* with statutory, police or other authorities concerned, *proposing or developing and* submitting to the Engineer for his consent, proposals based on such consultation showing a scheme of traffic safety and management measures including details of *safety zones* and emergency routes and furnishing such details as necessitated by the Works or as the Engineer may require;

(d) design of traffic safety and management measures specified by the Engineer;

(f) modification and resubmission of proposals and designs;

(g) traffic signs, *traffic signs provided by the Employer*, road markings, lamps, barriers, and traffic control signals including maintaining, cleaning, repositioning, covering, uncovering and removing;

(k) collecting and returning traffic signs provided by the Employer;
(l) surveillance and maintaining stocks;
(m) immediate reinstatement and replacement of defective or damaged items;
(n) maintenance of highways.

Box 22 MMHW — most introduced in Addendum

5 Damage to the Highway

The responsibility for repairing damage to highways rests with the Contractor unless stated otherwise in Appendix 1/17 or 1/18. The Conditions of Contract require the Contractor to insure and indemnify the Employer against loss, damage and claims and this is covered by Preamble 2(vii) to the Bill of Quantities.

Box 23 MMHW — Notes for Guidance

The Notes for Guidance set out the legal position and require Appendix 1/17 to show details where the Contractor in addition to routine maintenance functions 'is to be required when requested by a highway authority to repair accidental or wilful damage to any highway within the Site for which that authority is responsible'. The Sample Appendix does not specifically list this requirement and there is no obvious heading to cover it.

The specification clause does not separately identify this obligation of the Contractor unless it is considered to fall under 'If stated in Appendix 1/17, and to the extent there described the Contractor shall undertake the maintenance functions described therein on the lengths of highway there specified'. Responsibility for payment for this work would depend on Clause 20 (2) and (3) and particularly the excepted risks which include 'use or occupation by the Employer . . .'.

SHW Sub-clause 8 states that these maintenance functions do not relieve the Contractor of his obligations under Clause 22 of the ICE Conditions of Contract to indemnify the Employer for the Contractor's failure to maintain any highway adequately. The NG to MMHW state that damage to highways is the Contractor's responsibility (see Box 23) but this must be read in conjunction with the 'excepted risks' of ICE Clause 20 if the Employer or his agents are using part of the permanent works.

The Engineer should list in Appendix 1/17 'Any measures referred to in Chapter 8 and other Departmental documents which the Engineer requires the Contractor to carry out' (NG 117.5). This area was often a problem with the Sixth Edition as designers were reluctant to specify the detail required by the Notes for Guidance with resultant ambiguities regarding traffic arrangements and payment in respect of them. For the Contractor to price the tender properly it is vital that he is provided with the necessary detailed information through the contract documents.

The Contractor is still required to submit proposals for the Engineer's consent where the circumstances are not covered in the recommendations or described in Appendices 1/17 and 1/18.

If required in Appendix 1/17 the Contractor must appoint a Traffic Safety and Control Officer with one or more nominated deputies and arrange that one of them is on site at all times when work is proceeding.

In the event of an accident or breakdown occurring on a carriageway or hard shoulder the Contractor shall act on police requests subject to any instructions or contrary directions by the Engineer (such instructions will be covered by Clause 13 of the ICE Conditions of Contract enabling the Contractor to claim payment).

The table of Highways, Private Roads, and Other Ways Affected by the Works now includes a column specifying 'Whether to be Kept Open or Closed'.

If the Contractor is responsible for advanced signs, cones and road markings the areas of highway affected should form part of the Site and be included in Appendix 1/17. If he is not to be responsible, this should be stated in the appendix and any notice requirements for the moving of these signs shoud be stipulated so as to allow the Engineer to make the necessary arrangements.

The Details of Events That Could Have a Bearing on the Works is retained. SHW 117.2 requires the Contractor when planning the traffic safety and management measures to 'take into account the information contained in Appendices 1/17 and 1/18'. Care should be taken in completing this table as 'none' inserted against an item can lead to difficulties if this is not correct.

118 • *Temporary Diversions for Traffic*

This clause is substantially reworded and considerably clarified compared with the Sixth Edition. The Notes for Guidance and the layout of the Sample Appendix are similarly changed.

There were problems with the Sixth Edition over what constituted a temporary diversion for traffic with significant payment aspects. To avoid such problems the Model Contract Documents now define the term (see Box 24). Such clarification is welcome as this was one of the most widely misunderstood areas of the specification which resulted in many contract documents being wrongly drawn up. The effect of the new definition will be to exclude lane changes or single line traffic control from diversions. The Contractor will presumably now be expected to allow for these under the Traffic Safety and Management item. The MMHW has been significantly changed to identify separately diversions proposed by the Contractor and the NG to MMHW also seeks to clarify this area (see Box 25).

The only aspects of the Sixth Edition which are more or less unchanged are those in Sub-clauses 1 and 2 of the new clause which apply to all diversions (specified by the Engineer or proposed by the Contractor). These require that each temporary diversion shall be 'made operative in advance of any interference with existing arrangements,' maintained as provided for in the appendix and that the provisions of the clause do not apply to accesses, etc. solely for the use of the Contractor.

Separate divisions now apply to temporary diversions specified by the Engineer and those proposed by the Contractor. Where the Engineer specifies a temporary diversion (those 'required for reasons of safety or

'Temporary diversion for traffic' means:

(i) a temporary carriageway onto which vehicular traffic is diverted from the highway;
(ii) a temporary footpath or bridleway onto which pedestrian or equestrian traffic is diverted from a highway;
(iii) a combination of (i) and (ii) or a temporary carriageway as in (i) with an associated footway and/or way for the use of animals and equestrian traffic; or
(iv) a temporary private means of access onto which traffic is diverted from a private means of access;

but in all cases shall not include a central reserve crossover constructed to permit contraflow traffic on an existing carriageway.

Box 24 MCD — definition of temporary diversion

MMHW — Measurement

28 The measurement of taking measures for or construction, maintenance, removal of temporary diversion for traffic shall be in respect of the complete measures for or construction at the locations listed in Appendix 1/18 to the Specification and at locations proposed by the Contractor

29 New Group II feature

3 At locations proposed by the Contractor

Notes for Guidance on the MMHW — Series 100

4 Temporary Diversion for Traffic

The MMHW allows temporary diversions for traffic to be measured as follows:

(a) Specific Locations — These may include those where in the opinion of the Engineer the diversionary work is likely to be complicated, expensive, or its impact on or disruption of the Works is likely to be substantial. The description should include the appropriate reference from Appendix 1/18 of the Specification.

(b) Omnibus Item — This should include all diversions of a minor nature scheduled in Appendix 1/18 of the Specification.The single omnibus item should not include in its description the references from Appendix 1/18 of the Specification. A separate omnibus item should always be provided for all diversions proposed by the Contractor.

Box 25 Temporary Diversions — Measurement Aspects

practicality, including any structures' NG 118.2(i)), these must be detailed in the appendix, with construction, design and maintenance requirements, timescale for responsibility and any constraints which the Engineer should have agreed with the highway authority.

The Engineer must also ensure licences, etc. have been obtained, orders made and that the land over which the diversion is to be constructed has been made available and is included in Appendix 1/7.

The Contractor is then required to construct, maintain, remove and reinstate the diversions. If stipulated in Appendix 1/18 he is also required to design them to the details stated. The appendix may also stipulate where reinstatement of the ground to its original condition is not required.

The Engineer is required by the Notes for Guidance (NG 118.3) to provide full details if the Contractor is to be 'required to repair accidental or wilful damage to any temporary diversion of traffic specified by the Engineer at the request of the highway authority responsible for that diversion'. The Sample Appendix does not specifically list this requirement and there is no obviously appropriate heading to cover it. The specification clause does not separately identify this obligation of the Contractor unless it is considered to fall under 'shall be maintained to the standard stated in Appendix 1/18' (SHW 118.2). Presumably payment will be on the basis of the incurred cost or dayworks.

Outline proposals for any temporary diversion of traffic intended by the Contractor should be submitted for the Engineer's consent. The Notes for Guidance require the Engineer to 'only consent if the appropriate authority agree and should consult with the police before giving such consent.' The definition of diversions will limit this and will therefore exclude lane closures and single line working.

Any necessary statutory orders or notices required to be published should be submitted by the Contractor to the appropriate authority through the Engineer within the timescale stated in the appendix. (The Sample Appendix reference to SHW 118.6 is misleading — it should be to SHW 118.5).

The width of the temporary diversion 'shall be not less than that of the existing way unless otherwise agreed by the Engineer' and the Engineer should be informed of the Contractor's agreements with the highway authority for signs, lighting, construction, maintenance and removal.

Unlike the diversions specified by the Engineer, the Notes for Guidance do not advise the Engineer to set out in the appendix the responsibility of the Contractor for repairing accidental or wilful damage to a temporary diversion which he has proposed presumably because the Contractor will have agreed whether or not he is to do this work with the highway authority. Such a diversion will still be the responsibility of the highway authority as stated in the Notes for Guidance 117.4. The more general requirement of Appendix 1/17 will apply 'to any highway within the site' and the Engineer should consider this aspect when describing the details.

There does not seem to be any valid reason why the responsibility in respect of any temporary diversion specified by the Engineer is to be repeated in Appendix 1/18 if Appendix 1/17 already refers to it but NG 118.3 advises this.

As in SHW 117.8 this clause states that nothing in Sub-clauses 3 to 7 shall relieve the Contractor of his obligation under Clause 22 of the ICE Conditions of Contract to indemnify the Employer for the Contractor's failure adequately to maintain any highway (this must be read in conjunction with the Clause 22(1) (a)).

119 • Routeing of Vehicles

This clause has been simplified by the omission of the specific references in the Sixth Edition to routing of vehicles, movement of plant, use of the permanent works and temporary structures for the diversion of a public highway or to carry construction traffic. Where appropriate these are specified by the Engineer in Appendix 1/19 and the Contractor must comply.

The Sample Appendix covers permitted routes to the site (with details of any temporary traffic signs), movement of plant across public roads, use of the permanent works and extends the temporary structures to those over public highways and railways, rivers and canals.

The temporary structures specified by the Engineer and included in this appendix in the Sixth Edition is now omitted and should be included in Appendix 1/18.

120 • Recovery Vehicles for Breakdowns

This clause has been reworded with some additions to and omissions from the Sixth Edition. The main changes are the requirements for operators of recovery vehicles to hold a certificate showing successful completion of a course on breakdown recovery recognised by the Road Transport Industry Training Board and the proviso that operators accept instructions from the Engineer or police (subject to any contrary direction from the Engineer).

Minor changes include a clearer definition of 'roadworks operations', increased reference to Appendix 1/20, 'highway open to vehicles' rather than 'trafficked highway' and inclusion of abandoned vehicles (see Box 26 for changes to item coverage, some of which were introduced in the October 1991 Addendum).

There are no longer any references to AVRO when specifiying the recovery vehicles and Sample Appendix 1/20 is greatly expanded to cover the vehicle details and equipment. An additional requirement is that the Contractor should arrange for recovery vehicles to be inspected (SHW 120.1). Sample Appendix 1/20 gives details of this.

Details of the locations to which the vehicles are to be removed should be specified in Appendix 1/20 as should the locations for the recovery vehicles (if necessary with any specific requirements for hardstandings, etc.). NG 120.1(iii) is misleading in its reference to Sub-clause 120.6.

The restriction on the hours which the operators can work is omitted from the specification but NG 120.3 refers the Engineer to this aspect.

Recovery Vehicles 35 — Item Coverage

(a) equipment including *communication equipment and identification sign*;
(h) *qualified operatives*;
(i) completion and submission of *information log sheets*;
(o) *vehicle inspection and report*;
(p) *lighting board*.

Box 26 Changed Item Coverage

Box 27 MMHW — Changes to Preambles to B of Q — Largely included in Addendum

> Work Within and Below Non-tidal Open Water or Tidal Water
>
> 9 The Contractor shall allow in his rates and prices for taking measures required to execute the work separately measured as being within and below non-tidal open water or tidal water. *Subject to and without prejudice to the Conditions of Contract the datum stated in the Contract shall be used for the measurement of work affected by non-tidal open water or tidal water irrespective of the actual level of water encountered in the Works.*
>
> *Notwithstanding the foregoing, and the provisions of Chapter II paragraph 2, the Contractor shall allow in his prices for any items of structures designed by the Contractor provided under Series 2500, for taking measures required to execute work within and below non-tidal open water or tidal water although no separate measurement is provided and whether or not a datum is stated in the Contract.*

The Engineer and police are no longer specifically required by this clause to direct traffic onto an emergency route if the vehicle cannot be moved immediately as the Sixth Edition clause 120.11 is omitted but SHW 117.19 now covers this.

Additional requirements cover personnel wearing high visibility jackets (120.14), the communication system being operational before work which may require recovery vehicles commences (120.13), that recovery vehicle operators do not attempt to repair vehicles (120.9) and provision for a lighting board to be attached to the vehicle being towed (120.15).

The recovery vehicles must still be of a single recovery firm and once appointed they must not be changed without the Engineer's approval (SHW 120.1 and 120.4). There is no specific requirement for the Engineer's approval to the original appointment.

121 • Tidal, Flowing or Standing Water

The Contractor is to take measures to deal with tidal, flowing or standing water *within the site*. There are no Notes for Guidance but the Preambles to Bills of Quantities covers payment aspects (see Box 27). Care should be taken in specifying and measuring this work as the Contractor only allows in his rates for work separately measured as being within and below non-tidal open water or tidal water — a datum must be stated in the Contract. If the Notes for Guidance to the MMHW advice is followed in respect of the open water datum the Contractor may not have adequately allowed in his rates for the work likely to be encountered.

122 • Progress Photographs

Appendix 1/22 specifies the details — number, size, type and finish — of progress photographs. *The Notes for Guidance now omit references to the Bill of Quantity items and offer advice on photographs to cover operations which may give rise to third party claims.*

123 • *Use of Nuclear Gauges*

The Sixth Edition Clause 123A 'Blasting' is omitted and is covered in the 200 and 600 series.

The new clause sets out the requirements when the Contractor proposes to use, or is required in Appendix 1/5 to use, nuclear gauges and is intended to ensure that the Engineer and all site staff are aware of the possible dangers and regulations governing the use of this equipment. The Contractor must advise the Engineer of his designated 'radiation protection adviser' and provide a copy of his 'local rules'.

These matters are covered in the Ionizing Radiation Regulations 1985 and if the Engineer's staff are to use nuclear gauges a similar duty falls on the Engineer to inform the Contractor and notify particulars to the Health and Safety Executive.

The Contractor is not rendered immune from the relevant legal requirements by virtue of complying with this clause.

124 • *Substances Hazardous to Health*

This is a completely new clause setting out the procedure for, and restrictions on, the use of substances hazardous to health so as to protect the general public.

The Contractor has the primary responsibility for the execution of assessments under the Control of Substances Hazardous to Health Regulations 1988 (COSHH) and the development of safe working practices. There may, however, be instances where the use of substances allowed in the SHW could necessitate onerous restrictions on working methods to protect the general public or the environment. To enable the Contractor to allow for such restrictions in his tender, Clause 124 will be supplemented by Appendix 1/23 describing such restrictions.

The scheme designer will need to assess the use of materials which may represent a hazard and satisfy himself that the necessary measures to protect the public and environment are incuded in Appendix 1/23. It may be necessary to specify monitoring requirements if there is a possibility that hazards may occur under certain weather conditions and consider the need for suspending operations. The designer's assessment is simply a means to identify any requirements that should be brought to the attention of the Contractor.

Such substances may only be used when specified or with the consent of the Engineer. If the Contractor proposes to use a substance hazardous to health which has not been specified or proposes to use such a substance under the equivalence rules of Clause 104 he shall inform the Engineer of the measures he proposes to take to assess the risks and prevent, control and monitor such risk.

Copies of the Contractor's 'assessments of risk' and measures to protect those handling the substances (Regulations 6 and 7 of the Substances Hazardous to Health Regulations 1988) should be provided for the Engineer.

The Contractor must also provide written details of how he proposes to comply with Appendix 1/23 and for maintaining the control measures and monitoring the equipment. Appendix 1/23 supplements this clause and the Engineer should include any specific limitations on the Contractor's working methods when using such substances as silane, bridge deck waterproofing, paints, etc. These may include additional safety zones on traffic management measures, restrictions dependent upon wind speed/direction and when adjacent traffic speeds fall below a specified level, screening and signing and air quality monitoring.

If protective clothing is likely to be required by the Engineer it should be specified in Appendix 1/1 but SHW 124.7 requires the Contractor to provide it if necessary and to instruct the Engineer in its use.

Table NG 0/2 List of Sub-clauses which permit Contract-specific requirements to be included in the Contract instead of the national ones stated, e.g. Sub-clauses state '. . . unless otherwise described in Appendix -/-'

Preliminaries	
101.1	Removal of temporary accommodation and equipment for the Engineer.
101.2	Occupation date for accommodation for the Engineer.
103.2	Provision of separate communication system for the Engineer.
116.2	Interruption of privately and publicly owned services or supplies.
117.3	Traffic safety and management proposals.
117.4	Removal of central reserve cross-overs.
117.9	Provision of traffic signs.
117.10	Maintenance of traffic signs.
118.3	Removal of temporary diversions for traffic specified by the Engineer.
120.6	Limits of recovery service for breakdowns.

Table NG 0/3 List of Sub-clauses which require the Contractor to submit information to the Engineer.

Note: Information that the Contractor may submit when seeking the Engineer's approval is not listed in this table.

Preliminaries	
104.5	Quality management scheme or product certification scheme — submit certificates of conformity.
104.8	British Board of Agrément roads and bridges certificates — submit certificates.
104.10	Traffic signs requiring statutory type approval — submit written evidence of approval.
104.12	Departmental type approval/registration — submit type approval/registration certificates.
104.14	Provision of information — submit information and certificates.
104.15	Detailed working and fabrication drawings — submit drawings.
	Approved drawings — provide transparencies.
105.1	Suppliers of materials — submit list.
	Goods or materials — give details of choice.
105.2	Sampling and testing — supply test results and test certificates.
105.6	Sampling and testing — supply and deliver samples for test.
106.2	Structures designed by the Contractor — submit Approval in Principle form.
106.3	Structures designed by the Contractor — supply copies of completed design certificates.
106.4	Structural elements and other features designed by the Contractor — submit proposals.
111.1	Existing ground levels — submit schedule of levels considered to be in error.
112.1	Permanent ground markers and permanent bench marks — check and supply details.

112.2	Bench marks — keep updated schedules and supply when required.
113.1	Programme of Works — submit programme.
115.1	Accommodation works — give notice of start dates.
117.3	Traffic safety and management — prepare and submit traffic management proposals within given timescale.
117.5	Traffic safety and management — submit outline of central reserve cross-overs proposed by the Contractor.
117.6	Traffic safety and management — provide details agreed with the highway authority.
117.6	Traffic safety and management — submit application for statutory orders or notices.
117.9	Traffic safety and management — submit proposals for dealing with situations not provided for in the Contract.
117.11	Traffic safety and management — give notice within given timescale, for signs to be moved.
117.18	Traffic safety and management — provide details of the traffic safety and control officer and nominated deputes.
118.4	Temporary diversions for traffic proposed by the Contractor — submit outline of proposals.
118.5	Temporary diversions for traffic proposed by the Contractor — submit application for statutory orders or notices.
118.7	Temporary diversions for traffic proposed by the Contractor — provide details agreed with the highway authority.
120.12	Recovery vehicles for breakdowns — submit weekly logs.
122.3	Progress photographs — deliver within 4 weeks of exposure.
123.2	Use of nuclear gauges — advise name of the 'radiation protection adviser'.
	Use of nuclear gauges — provide copy of the 'local rules' for nuclear gauges.
124.3	— supply copy of assessment of the risks created and details of measures to be taken to prevent or control exposure.
124.5	— submit written proposals for implementing the requirements of Appendix 1/23.
124.6	— give details of measures proposed relating to any substance hazardous to health not specified in the Contract.
124.7	Substances hazardous to health — give details of training and health monitoring.

An Introduction to the Technical Changes

Before considering the changes it is worth noting that there are many areas where there has been little or no change. Thus the overall format of the documents has not changed and the title and content of individual series is little different. In most cases the clauses have retained their previous numbering and titles and, as before, the designer's input is through the various numbered Appendices. Although the arrangements for incorporating the Specification into the contract documents is broadly the same as in the Sixth Edition there are some significant differences.

Because of the loose-leaf format there is a need to control carefully the way in which the annual amendments are incorporated into the documents and this is achieved through a Schedule of Pages and Relevant Publication Dates. This is incorporated into the contract documents and lists the publication date of each page of the SHW applicable to the Contract — a standard schedule is included in the Notes for Guidance and a revised schedule will be published with each annual amendment to the Specification.

A standard Preamble is included in the NG 000 Series and must be reproduced and bound with the Numbered Appendices in the Contract documents. As well as setting out the general arrangement of the Specification it also states that in the event of conflict between the SHW and a numbered appendix, the numbered appendix will always prevail.

As with the Sixth Edition, Contract-specific requirements are incorporated through the numbered appendices with references in the clauses to 'as described/stated in Appendix */*' or 'unless othewise described in Appendix */*'. Tables NG 0/1 and NG 0/2 list the clauses containing such references.

The Specification should be used wherever possible as it stands and the proliferation of regional and other amendments that occurred with the Sixth Edition will be strongly resisted by DTp. HQ. Where alterations are necessary they need submitting to the DTp. in drafts of Appendices 0/1 and 0/2 with a justification. In no circumstances should Contract-specific amendments be incorporated by substitution of the loose leaf pages.

Where additional or substitute clauses are required or reference is made to a publication in a numbered appendix, the list of publications in Appendix F will need to be checked and any alterations included in Appendix 0/2.

The numbered appendices should contain the Contract-specific information, cross referenced to the drawings where necessary and should be based on the sample appendices with the content agreed by the DTp. They must only be used to extend the information in the SHW and never to change it. If new numbered appendices are required to extend Contract-specific alterations they should be numbered commencing at the 70th appendix of the series (e.g. 6/70) to avoid conflict with future national additional appendices.

Numbered Appendices 0/1 and 0/2 should contain Contract-specific alterations — additional, substitute and cancelled clauses in 0/1 and minor text changes in 0/2. In both cases if there are no alterations this should be indicated by 'NONE'.

Appendix 0/3 is to contain a complete list (A) of the numbered appendices referred to in the SHW with 'Not used' inserted as appropriate. List B contains any Contract-specific additional numbered appendices.

The **Construction Products Directive** is an EC Directive which is required to be enacted by national legislation. It is now in force and has two main effects — it is illegal to sell construction products which are not 'fit for purpose' and those products legally marketed must not be obstructed. **A CE mark** will signify that a product complies with the CPD and is, in effect a product passport.

The CPD sets out '**essential requirements**' which have been agreed at European level and to be effective it depends on 'technical specifications'. One of the main forms of these will be **Harmonised European Standards** which are based on the 'essential requirements' and as far as possible will be expressed in performance terms.

One of the primary methods of achieving the CE mark will be through compliance with a Harmonised European Standard transposed through a national standard.

Harmonised standards may cover only the relevant essential requirements but a **European Standard (EN)** may address matters over and above the essential requirements. European Standards must be implemented by equivalent national standards in member states.

CEN (Comité Européen de Normalisation) produces European Standards, generally by adopting international or national Standards in preference to writing new documents (e.g. EN 29000 which is equivalent to BS 5750). The EC has an agreement with CEN which allows the EC Commission to mandate CEN to produce Harmonised European Standards when it deems necessary.

Box 1 CPD and European Standards

Appendix 0/3 is to contain a complete list (A) of the numbered appendices referred to in the SHW with 'Not used' inserted as appropriate. List B contains any Contract-specific additional numbered appendices.

Appendix 0/4 is to list all of the Contract Drawings including those selected from the HCD.

National alterations in respect of Scotland, Wales or Northern Ireland are incorporated in Appendix 0/5 although it should be noted that the Preamble states that Numbered Appendices 0/1 and 0/2 take precedence over this appendix.

Tables NG 0/1 to NG 0/3 contain checklists of clauses which refer to numbered appendices, which allow for Contract-specific requirements and which require the Contractor to submit information to the Engineer.

Where British Standards are referred to in the SHW without a date the edition prior to the Reference Date will be applicable. (The Reference Date is defined in the Model Contract Document and would normally be three months prior to the return of tender).

Many of the clauses refer to British or European Standards and in an attempt to keep technically up to date those which are undated in the SHW are the respective editions current at the Reference Date which is defined in an amendment to Clause 1 of the ICE Conditions of Contract (normally three months prior to the date of the return of tenders). However, for those British Standards labelled with an asterisk in Appendix F the relevant issue is that stated in the Appendix. These are normally British Standards which have been amended or added to within the Specification or where a particular choice or range has been identified from those permitted in the BS.

The position is different when dealing with Harmonised European or European Standards (see Box 1). Where Harmonised European or European Standards have been issued prior to the Reference Date and since the last national amendment, Contract-specific alterations will have to be made to ensure they are properly incorporated into the Contract. SHW Clause 003.4 incorporates them by default but the designer will need to ensure this does not create any ambiguity or discrepancy in the documents.

Amendments to Harmonised European Standards are not permitted and amendments to European Standards should only be considered in exceptional circumstances.

Changes to reference documents after the Contract award can be made by the Engineer but if there are significant cost or delay implications then the DTp. should be consulted.

The Lettered Appendices have been retained although some have been relettered and retitled. Products requiring Statutory Approval before they are used on a public highway are now listed in a new Lettered Appendix D and products that require approval or registration by the DTp. before they can be used are detailed in new Lettered Appendix E.

The requirements of Motorway Communications have been transferred to the 1500 Series of the SHW with appropriate Notes for Guidance and MMHW.

The technical changes take account of the issued amendments to the Sixth Edition (Appendix L dated March 1988) and the Notes for Guidance (Amendment No. 1 dated March 1988), the amendments published as Departmental Standards or Chief Highway Engineer (CHE) Interim Amendments and other updates take account of technical improvements offering better value for money.

Due to the interdependence of the documentation the work on the new MMHW could not start until the Specification was largely complete and no further technical updating has therefore been carried out beyond December 1991. Thus the HMSO documents are dated December 1991 despite their later publication and work is already in hand on the first annual amendment.

The technical changes themselves do not represent a radical departure in technical content from the Sixth Edition although significant changes have been made to virtually every series. These changes are identified at the end of this introduction but some of the more significant technical changes are listed below under the headings of durability, speeding up construction, safety, minimising enviromental impact and collaborating with with Europe:

Durability: measures which are intended to lead to better long-term performance of products and materials.

Parapets, safety fences and lighting columns: anchorage requirements revised, welding requirements introduced.

Bridge bearings: requirements revised

Pavement drainage: requirements revised

Concrete pavements: dowel bar protection requirements revised.

Speeding up Construction: measures which permit earlier opening to traffic, with potential cost savings and benefits to road users.

Concrete pavements: use of rapid-hardening cement and new strength requirements permitting earlier opening to traffic.

Concrete pavements: laying of concrete permitted at low temperatures with use of insulation blankets.

Safety: measures designed to improve road safety and site safety during construction.

All safety fence drawings included in HCDs — most appropriate design can be chosen more easily.

Safety barriers: introduction of vertical concrete barriers and temporary concrete barriers as part of TM arrangements.

Health and safety: new requirements to safeguard site staff and public.

Minimising Environmental Impact: measures which make better use of local materials, or recycle what would otherwise be waste products, or improve the road environment through the use of materials which reduce noise.

Texture depth has upper limit to reduce traffic noise.

Earthworks: unacceptable material may be processed to be rendered acceptable.

Pavements: frost heave test improved to allow wider range of suitable materials.

Pavements: use of up to 10% recycled bituminous material is permitted.

Environmental barriers introduced: with new requirments for absorbent materials.

Collaborating with Europe: measures which are fundamental to the adoption of an 'EC friendly' specification, or are consequential on the need to remove potential barriers to trade.

Use of equivalent standards from other EC countries permitted.

Use of test results from accredited independent testing laboratories in other member states permitted.

Use of contractors' design of proprietary manufactured structures and structural elements introduced.

Type approval of bridge joints and registration of parapet anchorage systems introduced.

Summary of Significant Technical Changes

Series 200 — Site Clearance

1. Sealant treatment to lopped branches deleted.
2. Blasting for purposes other than site clearance (and earthworks excavation) now permitted.

Series 300 — Fencing and Environmental Barriers

1. Hedges deleted.
2. Accommodation works fencing included in permanent fencing.
3. Absorbent environmental barriers introduced.
4. Timber preservation requirements revised.

Series 400 — Safety Fence and Safety Barriers

1. Performance criteria added.
2. Welding requirements added.
3. Site testing of anchorages in drilled holes and of post foundations added.
4. Timber posts for UCB deleted.
5. Wire rope safety fence added.
6. Vertical concrete barrier and temporary VCB added.
7. New tensioning procedure introduced for TCB.

Series 500 — Drainage and Service Ducts

1. Requirements for combined drainage and kerb systems revised.
2. Flat invert corrugated UPVC pipes deleted but these and other types of pipe not included in Table 5/1 or 5/2 are permitted if a BBA Certificate is available.
3. Fin drains and narrow filter drains added.
4. Technical requirements for corrugated steel pipes added.

Series 600 — Earthworks

1. Provision made to require unacceptable material to be rendered acceptable.
2. Removal of surplus topsoil permitted subject to HQ agreement.
3. Earthworks tests, including frost heave, revised.
4. New grass seed mixture substituted.
5. Chalk for use as capping is now Class 6F2.
6. New Classes 6Q and 7H introduced for overlying fill to corrugated steel buried structures.

Series 700 — Road Pavements — General

1. Concrete or CBM permitted to be laid at temperatures below 3°C provided that thermal insulation blankets are used.
2. Clause for breaking up or perforating redundant pavement added.
3. Frost heave requirements revised (as Series 600).
4. Trafficking of concrete slabs, cement bound roadbase and sub-base related to the strength of the concrete.

Series 800 — Road Pavements — Unbound Materials

1. Soundness test introduced for aggregates of questionable durability.
2. Wet mix macadam deleted.

Series 900 — Road Pavements — Bituminous Bound Materials

1. Up to 10% reclaimed material allowed.
2. Specific requirements introduced for cleanness, hardness and durability of aggregates.
3. Clause for planing added.
4. TRRL mini texture meter deleted.
5. General revision consequent upon new editions of BS.

Series 1000 — Road Pavements — Concrete and CBM

1. Accelerated wear test deleted.
3. Bottom crack inducers required except where joint grooves are sawn. Transverse grooves to be sawn in summer.
4. Texture depth revised to 1.00 ± 0.25 mm
5. High early strength cements permitted

Series 1100 — Kerbs, Footways and Paved Areas

No substantial changes

Series 1200 — Traffic Signs

1. Statutory approval and authorisation formalised.
2. Detector loops added.
3. Class 1 retroreflectivity required for temporary signs.
4. Clause for traffic signals now gives more comprehensive requirements.

Series 1300 — Road Lighting Columns and Brackets

1. Site tests on anchorages in drilled holes introduced.
2. Welding requirements added

Series 1400 — Electrical Work for Road Lighting and Traffic Signs

1. Group switching of lighting circuits given greater prominence
2. Appendices reviewed. New Appendix 14/5 for electrical equipment for traffic signs.

Series 1500 — Motorway Communications

1. New series based on TCC Motorway Communications Manual.

2. Requirement for the contractor responsible for testing and terminating communication cable to be approved has been deleted.

Series 1600 — Piling and Diaphragm Walling

No substantial changes

Series 1700 — Structural Concrete

1. Surface impregnation added (based on BD 43/90).
2. Access requirements for cover measurement survey added.
3. Control of ASR and of durability update.
4. New Appendix 17/3 for contract-specific surface finishes.
5. Update BS for cements incorporated.

Series 1800 — Structural Steelwork

1. Revised to accord with European Standard BS EN 10025.

Series 1900 — Protection of Steelwork Against Corrosion

1. Use of sand for blast cleaning deleted.
2. Use of cadmium for metal coating deleted.
3. Alternative protective systems for steel lighting columns introduced.

Series 2000 — Waterproofing for Concete Structures

1. Requirement for waterproofing to be laid when temperature is above 4°C deleted.
2. Controls on heating of bitumen revised.

Series 2100 — Bridge Bearings

1. Steel chemical composition and hardness requirements revised for roller bearings.

Series 2200 — Parapets

1. Metal parapets to comply with BS 6779 as amended in SHW.
2. Welding requirements added.
3. Inspection and testing requirements for parapet posts introduced.
4. Site tests on anchorages in drilled holes introduced.
5. Anchorages and attachment systems to have Departmental registration.

Series 2300 — Bridge Expansion Joints and Sealing of Gaps

1. In situ nosings deleted (but guidance retained).
2. Bridge deck expansion joints to have Departmental type approval.

Series 2400 — Brickwork, Blockwork and Stonework

1. Control of ASR for mortar deleted.

Series 2500 — Special Structures

1. Technical requirements for corrugated steel buried structures reduced (reliance placed on Departmental type approval).
2. Pocket type reinforced brickwork retaining walls added.

Series 2600 — Miscellaneous

1. Control of ASR for bedding mortar deleted.
2. Maximum sulphate content for bedding mortar added.
3. Performance requirements for bedding mortar reviewed.

200 SERIES Site Clearance

In the following notes on individual clauses, the use of italics in headings, text and clause number denotes areas of change.

There have been minor rearrangements of the Specification, Notes for Guidance and Sample Appendices in this series with a general improvement in clarity, e.g. by the addition of 'by the Contractor' to SHW Clauses 201.8, 202.1 and 202.2. There are changes also in that the provision is now made for the preservation and protection of trees and shrubs (SHW 201.1) although the Sixth Edition requirement to seal the lopped ends of branches on preserved trees is omitted.

Clause 201.1 requires the Contractor to 'clear each part of the site at times . . . required or approved by the Engineer'. Sample Appendix 2/1 sets out the Engineer's requirements and mentions 'restrictions on when buildings, etc. can be demolished' which the Contractor should comply with. Any additional restrictions on timing would have to be clearly identified and if necessary shown in the Appendix to the Form of Tender as sectional completion. Any requirements on timing will have a bearing on the programme and should therefore be shown in Appendix 1/13. The sample appendix does not mention this (see Box 2 for new measurement features and changed item coverage).

The drawings should now show the 'area(s) included in the Bill of Quantities as general site clearance', rather than as previously 'the total area of the Site' (NG 201.3). See also Box 1 for drawing information to be provided.

> Obstructions above Ground Level
>
> The various Group I, Feature 3 items of site clearance measured separately are to be referenced on the site clearance drawings and listed in Appendix 2/1.
>
> The referencing of items for site clearance can include consolidated references such as 'a house with adjoining garage and outbuildings' providing that full identification is given in, or cross referenced in, Appendix 2/1.

Box 1 MMHW — Notes for Guidance

> Method of Measurement — Series 200: Site Clearance
>
> Unless otherwise stated in the Contract the items in this *Series* shall include for the removal of superficial obstructions down to existing ground level.
>
> *Lowering of carriageway levels shall be measured under Series 700 Pavements.*
>
> Para. 4 — Group I Feature 2
>
> General site clearance of separate sections.
>
> Para. 5 General Site Clearance Changed item coverage
>
> *(h) branch lopping;*
> *(j) reinstatement and making good;*
> *(k) preservation of individual or groups of trees, shrubs and the like;*
> *(l) treatment of hazardous materials.*
>
> Para. 9 New Group II Features:
>
> *1 Blockwork and stonework.*
> *3 Brickwork.*
> 4 Kerbs, channels, edgings, *combined drainage and kerb blocks.*
> *5 Copings, string courses and the like.*
> *6 Cable.*
> *8 Communications cabinets, posts, brackets, signal indicators and the like.*
> *9 Shelves, racking, frames and the like.*
> *10 Electronic units and the like.*
> *12 Individual blocks, features or stones.*
>
> Para. 9 New Group IV Feature I *Different arrangements*

Box 2 MMHW — New Features and changed item coverage

Buildings are part of site clearance down to ground level and a schedule should be included in Appendix 2/1. *Parts below ground level which are to be removed 'as an earthworks operation' (NG 201.1) should be indicated on the drawings and where necessary measured as 'hard material'. The previous Notes for Guidance indicated that parts of buildings below ground level would be removed as site clearance. This reference is now omitted.*

Holes are to be made over at least 10% of the area of slabs, basements, etc., liable to hold water. If required this should be clarified in Appendix 2/1 as the item coverage does not specifically mention this.

The Contractor should take all measures required by any Statutory Undertaker (*SHW 201.5*) for proper sealing off, etc. Item coverage includes 'disconnecting, removing and sealing services and supplies' but it will often not be possible to define this accurately in advance. The Engineer should take care not to approve any measures without consulting the Statutory Undertakers. The Contractor's responsibilities are qualified by the MMHW General Directions.

Materials become the property of the Contractor unless included in Appendix 2/3 or required for use in the Permanent Works. Material to be recovered should be listed in Appendix 2/3 — damaged items are to be replaced (see Box 3).

Damage to Items

Item coverage includes for replacing items damaged in the process of taking up or down and setting aside or storing. It is the Contractor's responsibility to ascertain at the time of tender the extent of any damage which may occur and to make the appropriate allowance in his rates and prices.

Box 3 MMHW — Notes for Guidance

Topsoil for parapet walls should be reserved for re-use and spread over disturbed ground — any surplus disposed of 'as described in Clause 602'. This cross-reference to the 600 series leads to problems, as Clause 602 requires topsoil Class 5A to be stockpiled rather than disposed of (see the notes on Clause 602) and the MMHW measurement rules do not appear to take the site clearance topsoil into account. The item coverage for general site clearance merely refers to 'disposal of material'.

Voids are to be backfilled immediately when specified in Appendix 2/3 (201.6) and holes left by tree stumps filled within one week (202.2).

Blasting for site clearance was only allowed in the Sixth Edition if Appendix 2/4 permitted it, now it is 'to be confined to the locations and be within the limits stated therein'. Compliance with Clause 607 is required except that references to Appendix 6/3 are replaced by references to Appendix 2/4. Other blasting can only be employed with the consent of the Engineer.

The treatment of hazardous material 'shall comply with any specific requirements stated in Appendix 2/5'; the 'or as agreed by the Engineer' is now omitted. However, the wording of the Notes for Guidance and Sample Appendix 2/5 require only guidelines of what would be acceptable to the Local Authority — the HSE are mentioned in NG 204 but not in the sample appendix.

Additionally it is now stated in SHW 204.2 'Compliance with sub-clause 1 of this Clause does not confer immunity from relevant legal requirements'.

Table NG 0/2 List of Sub-clauses which permit Contract-specific requirements to be included in the Contract instead of the national ones stated, e.g. Sub-clauses state '. . . unless otherwise described in Appendix -/-'

Site clearance	
201.2	Severance of lines of existing fences, hedges or walls.
201.4	Removal of disused services.
201.4	Backfilling trenches.
202.1	Disposal of felled timber.
202.2	Removal of stumps and tree roots.

300 SERIES Fencing and Environmental Barriers

Introduction

A new 'general' clause and rewording of other clauses removes many of the difficulties of the previous specification regarding the requirements for accommodation fencing.

There is an increased reliance on the HCD drawings and on Appendix 3/1 but with surprisingly little change to the wording of some of the Specification and Notes for Guidance clauses.

Environmental barriers (previously noise barriers) will normally be for the Contractor to design and should therefore be listed in Appendix 1/10 and he must submit his design for 'aesthetic approval'. There are also new requirements for absorbant materials.

Hedges have been removed from this series and no longer appear in the specification

In the following notes on individual clauses, the use of italics in headings, text and clause number denotes areas of change.

301 • General

This is a new clause which requires 'Temporary fences, permanent fences and environmental barriers' to comply with this series. This simple clarification is welcome as the Sixth Edition requirements for temporary fencing and accommodation fencing were very poorly worded in this respect with substantial ambiguities.

302 • Requirements for Temporary and Permanent Fences

The Contractor must 'immediately' erect fencing. This needs care in interpretation as it will often not be practical. Clause 115 also requires the Contractor to give at least ten days' notice of the start of any accommodation works and this will often apply to fencing.

Temporary fencing is required if permanent fencing is not erected immediately or where none is required. As the Contractor must allow in his rates for temporary fencing the Engineer's interpretation of 'immediately' has considerable payment consequences (see Box 1 for measurement aspects).

On the 'boundaries of the land' will not always be correct (SHW 302.1). This should be clarified on the drawings.

The type of temporary fencing is still 'to be chosen by the Contractor'

Temporary Fencing (Notes for Guidance)

The Specification requires the Contractor to erect temporary fencing in all situations where he does not provide permanent fencing immediately. To comply with the Specification, Health and Safety Regulations and the Conditions of Contract the Contractor has the choice of a range of four specified types of temporary fencing. This temporary fencing is not shown on the Drawings nor is it included in the Bill of Quantities. However, should some specific temporary fencing be required by the Engineer then this should be shown on the Drawings and included in the Bill of Quantities.

Temporary Fencing — MMHW

Temporary fencing required by the Contractor in the discharge of his obligations under Clauses 19 and 22 of the Conditions of Contract shall not be measured.

Box 1 MMHW and Notes for Guidance

but it now 'may be selected from the four standard types for highway works, etc.' Is this 'may' meant to be a 'shall'? (See also SHW 303).

Any specific requirements for temporary fencing should be shown on the drawings, e.g. right of way, but the Notes for Guidance now omit the 'and appropriate items included in the Bill of Quantities' (see Box 1).

The requirement for concrete footings should be specified in Appendix 3/1 — HCD show details. MMHW requires a separate item for concrete footing to fence posts (see Box 2).

MMHW still includes excavation in hard material in item coverage although the wording is changed from 'of' to 'in'. This needs to be read in conjunction with the basic principles of the MMHW (see Box 3).

Concrete footing to intermediate posts shall only be measured for those locations stated in the Contract.

Box 2 MMHW — Measurement

303 • Temporary Fencing

The clause now particularly requires the temporary fencing to 'be appropriate to the usage of the adjoining land' but the previous clear requirement that it 'shall be any of the following types' is changed to 'may be selected from the following types' — and presumably also it may not (see Box 1 for measurement aspects).

A qualification 'subject and without prejudice to the Conditions of Contract' has been added to SHW 303.2 and a new Sub-clause states that the timber need not have preservation treatment (unless required in Appendix 3/1).

304 • Timber Quality

Changes to this clause include minor rewording, the replacement of Table 3/1 with HCD Drawings H37 and H38 and the omission of references to timber for safety fences.

It is important to recognise and ensure compliance on timber species. Spruce posts have often been used in the past in contravention of the Specification. There is considerable emphasis on British Standards and structural timber must be stress graded.

Compliance with this clause before preservation treatment is a requirement of Clause 311.1 with considerable implications in terms of supervision. *Samples shall be taken by the manufacturer or supplier in accordance with the sampling method in BS 1722:Part 7. Note that different rejection numbers are used for environmental barriers (there is no reference to Appendix 1/6).*

305 • Fittings

Minor changes have been made to include 'screws and nuts' and as black bolts grade 4.6 are no longer generally available, cold forged bolts grade 4.8 are allowed as an alternative.

306 • Permanent Fencing

With the rewording of SHW Clause 307 and the new General Clause 301 it is now clear that the requirements of this clause also apply to

Hard Material

Excavation in Hard Material occurs in the item coverage for several items of work, for example, fencing, safety fencing, traffic signs and road markings, road lighting and electrical work. The Contract should contain information known to the compiler about the existence and extent of Hard Material and this should include existing buried roads and the like. This would not relieve the Contractor of his obligations under Clause 11 of the Conditions of Contract. Hard Material is measured extra over normal excavation for earthworks and drainage and guidance is given under Series 600.

Box 3 MMHW — General Principles — Notes for Guidance

Group I	
1	Each type of fencing
2	Concrete footing to intermediate posts of each type of fencing.
3	Each type of gate.
4	Each type of stile.
Group II	
1	Fencing of different heights.
2	Gates of different heights and widths.
Group III	
1	Painted fencing, gates or stiles.

Box 4 Itemisation — Fencing, Gates and Stiles

accommodation works fencing, thus removing one of the difficulties of the Sixth Edition. The wording of Sub-clause 1 would imply that Appendix 3/1 will now include accommodation works fencing but it seems as though this is to be shown in Appendix 1/5 — an area of potential confusion (see Box 4 for MMHW itemisation).

Permanent fencing is required to be of a flowing alignment in plan and elevation (unless otherwise described in Appendix 3/1). This needs careful consideration at design stage if strained wire is used (NG.306.5) and may affect landtake. If trimming under fence is required this should be stated in *Appendix 3/1 with the location* — item coverage includes 'trimming ground on line of the fencing' (see Box 5 for MMHW guidance and item coverage).

Change in fence type should be at an appropriate feature so as to give a logical reason for the change. Only two fencing types are chosen to encourage standardisation — one wooden and one dropper — which should be sufficient to meet 'all environmental and amenity factors'. Any additional stockproofing required should be specified in Appendix 3/1 (NG 306.4 offers suggestions) together with any requirement for painting. Wire dropper fences for motorways are to comply with BS1722:Part 3: 1986 Sections 4 and 5 but with many detailed alterations.

Fences joining to existing hedges, fences, etc., are to be as HCD.

The Notes for Guidance are now reworded and it is clear that the drawings should show the position, length and height of each type with Appendix 3/1 giving any further details. A new Notes for Guidance Sub-clause 8 requires the Engineer to specify in Appendix 3/1 if concrete surround is required to posts. The Sample Appendix includes this.

> Notes for Guidance
>
> Where the Engineer considers that the ground conditions necessitate either concrete foundations to posts or longer posts the details and locations should be shown in the Contract. In the case of longer posts they should be given a specific type reference.
>
> MMHW — Item Coverage
>
> (d) trimming ground on the line of the fencing;
>
> (k) adjustment of fencing to a flowing alignment including additional length posts;

Box 5 MMHW and Notes for Guidance

307 • *Permanent Fencing for Accommodation Works*

The wording of this Clause in the Sixth Edition left it open to question as to what extent Clauses 304, 306 and 311 applied. This uncertainty was further compounded by the specification clause referring only to Appendix 1/15 whereas Clause 306.1 implied that 'all permanent fencing' was described in Appendix 3/2. This ambiguity was commonly resolved by the designer specifying in Appendix 1/15 that the requirements of Series 300 applied. *This is no longer necessary as the new wording and the new title of the clause makes the position clear but confusion may arise as the wording of SHW 306 and 307 imply details of accommodation works are included in different appendices.*

Landowners should be encouraged to choose HCD fencing type.

308 • Gates and Stiles

The clause now refers to various HCD drawings for details and dimensions rather than 'as described in the Contract' (see Box 4 for MMHW itemisation).

Para 17

(c) excavation *in* Hard Material (as Series 600 paragraph 23);
(d) trimming ground on the line of the *environmental* barrier
(l) sample panels including *safety fences*;
(m) gates, fittings and *furniture*;
(n) adjustment of the *environmental* barrier to a flowing alignment;
(p) *forming pockets and casting in sockets*;
(r) fixing to structures and *other elements*;
(s) *accesses*.

Box 6 Environmental barriers — changed item coverage

309 • Removing and Re-erecting Existing Fences and Gates

There is only minor rewording in this clause.

Take down is measured under site clearance and re-erected under fencing.

310 • *Environmental Barriers*

The Sixth Edition reference to 'Noise Barriers' has been changed to 'Environmental Barriers'.

Additional width may be required on embankments to accommodate panels behind gaps (*NG 310.2*). Where safety fencing should be provided between the carriageway and the environmental barrier or where space is limited it may be integrated with the barrier.

Materials specified should be chosen after consultation to ensure they are suitable (NG 310.7) and climbing plants may be included.

Aesthetically, alignment should be flowing and steps avoided (tolerances 75 mm horizontally and 50 mm vertically). This implies very careful specifying of the ground to be excavated or filled and compacted. Item coverage mentions trimming but not filling or compaction. *Unless stated otherwise in Appendix 3/2 the barriers should be vertical with no stipulated tolerances.*

Normally the Contractor should design the barriers and these should be listed in Appendix 1/10 together with full requirements in Appendix 3/3, including whether he is to design the foundations, etc. (NG 310.5) (see Box 7 for measurement rules in this case).

There could be some confusion over who gives aesthetic approval — the Contractor is now required to 'submit his design to the Engineer for aesthetic approval' (SHW 310.14). However, NG 310.12 states that Appendix 3/2 should include factors 'which the Contractor will need to take into account when formulating his submission for Departmental aesthetic approval' and also states the Engineer submits the proposals 'to the Overseeing Department for aesthetic approval'. The intention is that aesthetic approval is given by the Overseeing Department and SD4/92 requires the Engineer to indicate the timescale as constraints in Appendix 1/13.

Environmental Barriers

When designed by the Contractor these barriers are to be measured under Series 2500.

Box 7 MMHW — Notes for Guidance

The Engineer is required to check the Contractor's design and various points are suggested in NG 610.11 for the Engineer to consider. These do not include the aesthetics.

The Engineer's position with regard to responsibility for the Contractor's design is set out in an additional sub-clause added to ICE Conditions of Contract Clause 8 through the Model Contract Documents. This states 'The Engineer shall accept responsibility for the Contractor's design after the check procedures and shall notify the Contractor of the date of acceptance'.

If sample panels are required by *Appendix 3/2* they should be erected not less than six weeks before starting construction for approval by Engineer. There is no period stated for the Engineer's approval or procedure in the event of him not approving them or requiring changes, nor is there a definition of 'construction'. NG 310.10 refers to 'mass production'.

If design drawings are required these should be specified in Appendix 1/7. *Additions to this clause include 'Other materials' (Sub-clause 13) and extensive requirements covering testing, absorbtion and insulation of 'Absorbent Barriers' Sub-clauses 15–19.*

311 • Preservation of Timber

This clause has minor rewording only but with a significant change in Sub-clause 2(iii) where the Engineer can now accept timber with moderately resistant sapwood which has not been completely penetrated by preservative.

The performance standard chosen is based on 40 years life and is intended to keep maintenance costs to a minimum. Either creosote or CCA is permitted and if one of these is to be specified this must be made clear in Appendix 3/1.

NG 311.1 no longer 'recommends' but simply states that the same standard of preservation treatment applies to accommodation works. The savings in maintenance will therefore be to the benefit of the landowner rather than the DTp.

Timber quality must comply before treatment. NG 311.2 stresses this is of the utmost importance and requires it to be carried out for each scheme. This implies extra supervision and as this is an accepted Quality Management Scheme operation (see extracts from Appendix A at end of this section), it appears contrary to QA philosophy.

NG 311.6 also stresses the need to ensure moisture content does not exceed 28% prior to treatment and states that 'The only way to ensure that the preservation treatment has been carried out as required is to witness it'. These very strict requirements affect the Engineer and could have serious consequences if he does not comply with the advice of the Employer.

The testing of the moisture content lacks clarity — SHW 311.2(v) states 'Where required by the Engineer, moisture content determination, using a moisture meter as described in BS 5589: Section 6: *1989* shall be confirmed using the oven drying method, etc.'. The relevance of the confirmatory tests is not clear from the Specification even though the criteria for the test are set out in the clause. This could be overcome if the designer specifies in Appendix 3/1 what is required — the Sample Appendix does not cover this.

Group III Feature 1
Painted fencing, gates or stiles.

Box 8 Itemisation — Fencing, Gates and Stiles

Batch delivery to site should be intact and accompanied by a certificate with details of species, treatment, moisture content, etc. This does not seem to be consistent with the Notes for Guidance requirement for witnessing the treatment.

Preservation plant needs to be carefully assessed — NG 311.11 gives indications as to quality and amount of equipment to be expected.

The strict criteria for compliance in preservation and timber quality have serious implications for the RE's staff in supervising, the Contractor in pricing and the timber suppliers in complying.

312 • Painting of Timber Fences, Gates, Stiles and Posts

This clause has only minor rewording.

Colour of paint and whether preservation treatment is required should be described in the Contract. Alternatives to oil based paint can be specified if required. If the fencing is accessible to animals the paint should be specified as non-toxic.

313 • *Pedestrian Guard Rails*

There is no longer a Clause 313 — these requirements are now covered in clause 412

Table NG 0/2 List of Sub-clauses which permit Contract-specific requirements to be included in the Contract instead of the national ones stated, e.g. Sub-clauses state '. . . unless otherwise described in Appendix -/-'

302.1	Use of barbed wire on temporary fencing.
302.2	Removal of temporary fencing.
303.1	Type of temporary fencing.
303.3	Preservation treatment of temporary fencing.
304.2	Timber quality.
305.1	Grade of bolts, screws and nuts.
306.1	Alignment of fencing and joining to existing hedges, fences etc.
310.7	Surface preparation of timber.
310.14	(i) Barrier post spacing.
310.14	(vi) Stepping of panels.
311.2	(i) Preservation of timber.

Table NG 0/3 List of Sub-clauses which require the Contractor to submit information to the Engineer.

Note. Information that the Contractor may submit when seeking the Engineer's approval is not listed in this table.

310.14	Environmental barriers designed by the Contractor — submit design for aesthetic approval.
310.14	(i) Sample panels — obtain Engineer's approval.
310.14	(iii) Environmental barriers designed by the contractor — supply drawings, design calculations and details.

APPENDIX A Quality Management Schemes

2 Description: Supply and Erection of Fences (QGN 5000/5, formerly QAS 4630/5)

Certification Body: BSI Quality Assurance
PO Box 375
Milton Keynes
MK14 6LL

Specification: General fencing shall be in accordance with the 300 Series of the Specification and the following British Standard: BS 1722: Specification for fences Parts 1 to 13.

5 Description: Timber Preservation (QAS 4620/290 and allied schedules)

Certification Bodies: BSI Quality Assurance
PO Box 375
Milton Keynes
MK14 6LL

TRADA Quality Assurance Services Ltd
Stocking Lane
Hughenden Valley
High Wycombe
Buckinghamshire
HP14 4NR

Specification: Chemicals used and timber preservative treatment to be carried out shall comply with Clause 311 of the Specification.

Note: Timber used for general fencing shall be treated by firms registered by these bodies, and firms shall only obtain chemical specialities used in treatment from a registered source.

400 SERIES — *Safety Fences, Safety Barriers and Pedestrian Guardrails*

Introduction

The 400 Series has been very extensively rewritten to include double-rail single-sided open box beam, wire rope safety fence, concrete safety barriers and pedestrian guardrails.

Performance criteria have been introduced for safety fences and barriers based on full scale dynamic testing relating to their ability to contain and redirect vehicles safely. The lateral deflection measured during the test will be a criterion determining where that type of fence may be used since the clearance between the fence and an obstacle should not exceed the measured deflection. Durability requirements ranging from 10 to 50 years are also introduced.

The specification for materials and fabrication has been updated and it is intended that further updating will take place to keep pace with European Standards for steel that are currently being finalised and published at frequent intervals. Some current standards are in need of updating and these — for example BS 4190 (1967) — have been suitably amended in the specification.

New requirements cover components including welding and testing and the Contractor must provide evidence regarding the ability of anchorages to resist the specified design stresses.

A new clause sets out the detailed testing of post foundations and the item coverage has been extended to cover this rather than a separate item being included in the bill.

There are extensive new clauses to cover double-rail single-sided open box beam (DROBB), wire rope safety fence (WR), and concrete safety barriers.

All safety fence drawings are now included in the HCD Section 2 (approximately 200 drawings) so that the most appropriate design can be chosen more easily and they are also on CAD. Previously drawings were issued by TRRL and others.

Pedestrian guardrails, previously in the 300 Series are now included in this series.

The MMHW requires that if the Contractor is to be given the choice of safety fence type (WR or TCB) separate bills of quantities are to be provided and provision made for only one bill to be carried to the tender total. The MMHW has also been extended to cover the new types of safety fence and barrier and the item coverage has been considerably revised. The October 1991 Addendum forms the basis for much of this change but often in a modified form (see Boxes 1 and 4 for information on billing, Box 2 for changed item coverage and Box 3 for inconsistent item coverage).

> In the following notes on individual clauses, the use of italics in headings, text and clause number denotes areas of change.

401 • Performance Criteria for Safety Fences and Safety Barriers

Safety fences and barriers must comply with the specified criteria in respect of their ability to contain and redirect vehicles so that after impact they are contained and redirected onto a path adjacent to the barrier. The degrees of containment are Lower, Normal and Higher dependent on the mass of the vehicle, its speed and its angle of incidence. The normal level relates to a 1.5 tonne vehicle travelling at 70 m.p.h. at an angle of 20° to the fence.

Alternative Types of Safety Fence

6 Where the Contract provides for the tenderer to select either wire rope safety fence or tensioned corrugated beam safety fence to be constructed over given lengths, separate Bills of Quantities are to be provided containing alternative types of safety fences for the lengths in question. These Bills of Quantities are to be included within Series 400: Safety Fences, Safety Barriers and Pedestrian Guardrails.

Immediately preceding the separate alternative Bills in Series 400 an Index (Table 3) is to be provided for the alternative types of safety fence permitted by the Contract.

Provision is to be made for only the one Bill of Quantities in Series 400 of the Roadworks General Bill which relates to the type of safety fence elected to be constructed by the Contractor to be priced and included in the Tender Total.

Box 1 MMHW Chapter III

Durability requirements are a design life of 20 years for safety fencing, 50 years for concrete barriers and 10 years for temporary vertical concrete barriers.

The safety fences and barriers described in the 400 Series are deemed to meet these criteria and have received Departmental type approval. The Engineer will not require a copy of the certificate if these types are used. Other fences and barriers are permitted but must comply with those criteria and with Clause 104 (see Box 1 for information on billing).

402 • Components for Safety Fences and Safety Barriers

This new and extensive clause covers the requirements in respect of materials (steel main components, S & Z posts, fittings and concrete); protective finishes; tolerances; extensive requirements for welding including inspection and testing; marking; workmanship and testing, handling and storage (manufacture of safety fence components is an accepted Quality Management Scheme in Appendix A).

The contract drawings should include the relevant drawings from the HCD: Section 2 for the safety fences or barriers scheduled in Appendix 4/1 together with the type of intermediate post foundation based on the Engineer's selection — wherever possible using tested ground conditions (NG 402.1 and 2).

BS 4190 (1967) is regarded as restrictive in the light of modern practice and amendments have been introduced through SHW 402.2(iii). These include changes to grade 4.8 (cold heading process) black bolts as the BS grade 4.6 (hot forged) bolts are no longer generally available.

Generally, components must be galvanised in accordance with Clause 1911 with the exceptions stated in SHW 402.3. BS 729 (1967) is in need of updating and the specification has amended it in respect of pickling prior to galvanising and the requirements are changed in respect of thickness of steel at which the coating weight changes. This is necessary as many components are fabricated from steel of 5 mm nominal thickness — the BS point of change. The change is to a thickness of 5.5 mm.

Comprehensive new requirements have been added covering welding procedures, welder qualification, production inspection and testing, destructive testing and remedial work. Written welding procedures are required to BS 4870 and these are subject to re-approval every seven years. Welders must be qualified in accordance with BS 4871 and are subject to re-approval every two years. The manufacturer organises this approval and also provides personnel to carry out visual inspection and non-destructive testing of production welds.

Welding should only be used where detailed in the drawings and the Engineer should request copies of certified destructive test reports prior

Posts — Para. 10 — Item coverage

(a) *fabrication (as Series 1800 paragraph 6);*
(b) protective system (as Series 1900 paragraph 4);
(c) *driving in any material;*
(d) fixing to structures including *attachment systems;*
(e) fixing to beam including spacers;
(f) facilities for Engineer's *proof loading;*
(g) *drilling or forming holes and pockets and casting in bolts, base plates and anchorage assemblies;*
(h) bedding;
(i) filling;
(j) proof loading.

Box 2 MMHW Posts — some changes to item coverage — some introduced in Addendum

to delivery of components. He should arrange for sample components to be selected by a welding inspector and mark and dispatch them for destructive testing at an accredited NAMAS laboratory (NG 402.3, 4 and 5).

The frequency of sampling is covered by SHW 402.5(v) — 1 sample joint per 100 beam assemblies; 1 driven post per sampling lot not exceeding 1000; 1 surface mounted post per sample lot not exceeding 100; and anchor frame at six-month intervals (there is no reference to Appendix 1/6).

The criteria for selecting the samples is covered by the Notes for Guidance. These should take account of manufacturers' inspection reports, previous test results and production practice — 'where practicable, samples should be selected on the basis that they represent the lower end of quality in the production batch' (NG 402.4).

Notes for Guidance Clause 402.6 supplements the SHW information by setting out points which 'The Engineer should consider ... when ascertaining the acceptability of components subject to destructive testing'. The references in NG 402.6(a) to the SHW clauses are incorrect in that 402.4 (iv) and v(a) do not exist — presumably they are intended to refer to 402.5 iv(a).

If non-conformance occurs action depends on how serious the deviation is and on its traceability.

Test reports and samples are to be returned to the manufacturer and be retained and made available on future contracts. Components should be clearly marked in accordance with Clause 402.6. Testing requirements for wire ropes/terminals are set out in Clause 402.7(iii) and (iv).

403 • Installation of Safety Fences

The Overseeing Department should be consulted when deciding on whether to specify safety fences or barriers.

HCD drawings show general requirements. Any particular requirements for installation should be described in Appendix 4/1 together with locations, etc. Wherever appropriate the Contractor should be offered a choice of tensioned beam or wire rope safety fence (see Box 1).

The requirement for the fence to be to a flowing alignment is still retained with tolerances on plan as in the Sixth Edition. The poorly worded 'deviate in 10 m' is unchanged. The tolerances on height in the Sixth Edition are omitted. New requirements appear in respect of excavation for concrete foundations and anchor blocks (SHW 403.3) and for the concrete (SHW 403.4).

The cumulative length tolerance is such that beams and posts can be positioned within ±30 mm of their prescribed location. The very strict and impractical requirement of the Sixth Edition Clause 402.4 that 'Each post shall be located within ±12 mm of its prescribed distance from the datum' is omitted (see Boxes 2 and 3 regarding measurement aspects).

If posts are mounted in cast-in sockets they shall be filled with non-setting

Concrete Foundations to Posts — Para. 36 — Item coverage

(a) excavation in any material (as Series 600 paragraphs 17, 18, 19, and 23);

Box 3 Compare with Para. 13 (a), (b) and (c)

Itemisation

8 Separate items shall be provided for beam safety fences in accordance with Chapter II paragraphs 3 and 4 and the following:

Group	Feature	
I	1	Each type of beam.
	2	Each type of post.
	3	Each type of mounting bracket.
	4	Each type of terminal *section*.
	5	Each type of full height anchorage.
	6	Each type of expansion joint anchorage.
	7	Each type of connection to bridge parapet.
	8	Each type of *connection* piece.
	9	Each type of concrete *foundation* to post.
	10	*Each type of socketed foundation to post.*
II	*1*	*Straight or curved exceeding 120 metres radius.*
	2	*Curved exceeding 50 metres radius but not exceeding 120 metres radius.*
	3	Curved not exceeding 50 metres radius.
III	*1*	*Double rail.*
IV	*1*	*Double sided.*

MMHW — Notes for Guidance

The MMHW provides for three categories of curvature for payment purposes. Curves which are made up from individual straight lengths of beams should not be considered to be small lengths of straight fence. They should be measured as curved fences within the Group II Features in MMHW. The radius is to be considered to be the radius equal to that of the arc which passes through the posts.

Box 4 MMHW Para 8 and Notes for Guidance

passive filler after erection if required by Appendix 4/1. This filler should be easily removable (NG 403.8).

At least four weeks (can be changed by Appendix 4/1) before installation of surface mounted posts the Contractor must provide 'well attested and documented evidence' regarding the capability of the anchorages to resist the design stresses given in Table 4/3. NG 403.4 gives an indication of what testing/certification would be acceptable and the manufacturer should supply details of maximum tolerances in the size of the hole permitted.

404 • Site Testing

Appendix 4/1 should describe whether the Contractor or the Engineer will provide equipment and carry out the tests on post foundations. Details of the test equipment are given in the HCD Section 2. Tests should be carried out at a time when the ground has least resistance (see Box 2 for changes in item coverage with regard to testing).

If the Contractor is carrying out the tests he must provide the results to the Engineer 'at least one week prior to installation of relevant lengths of fence, unless otherwise stated in Appendix 4/1'. If the Engineer provides the equipment and carries out the test the Contractor must provide a five-tonne vehicle and install and remove test posts and their foundations.

Where appropriate the Contractor must 'establish and maintain traffic safety and management measures complying with Clause 117 during installation, loading and removal of the test posts and foundations'.

405 • Tensioned Corrugated Beam Safety Fence (TCB)

This clause specifies only the assembly and tensioning requirements with components, installation and testing being described in the previous clauses. The Sixth Edition (Clause 402) requirements are considerably changed with all the sub-clauses being reworded.

The importance of the removal of longitudinal clearance is still emphasised but the method of achieving this is now specified — 'prizing apart the beams using a tapered bar in the 20 mm diameter hole of one beam and the end of the other beam'.

Surprisingly the impractical requirements of the Sixth Edition are retained in SHW 405.4(iii) — 'Anchor bolts and nuts at both anchor positions shall be tightened until the adjacent posts just maintain their vertical alignment without signs of movement'.

The Engineer must be satisfied on the security of the anchorage before tensioning proceeds (see Box 5 for changes to item coverage for anchorages). The procedure for tensioning is changed in that a torque of 280/300 Nm is first applied (to remove residual slack in the fence) and then released. Tension is then re-applied at a torque appropriate to the ambient temperature. There is a greater range of temperatures during which tensioning can take place (22 °C to −5 °C compared with 10 °C to 20 °C). Table 4/4 gives information on this.

A new requirement regarding the position of the securing screw is detailed in SHW Clause 405.5. Many safety fences already erected under the Sixth Edition would fail on this aspect.

Terminal Sections, Full Height Anchorages, Anchorages, Connections to Bridge Parapets and Connection Pieces

- *(a)* *Posts (as this Series paragraph 10);*
- *(b)* *beams (as this Series paragraph 9);*
- *(c)* *excavation in any material (as Series 600 paragraphs 17, 18, 19, and 23);*
- *(d)* *concrete (as Series 1700 paragraphs 4 and 9);*
- *(j)* *precast concrete fairings;*
- *(l)* *casings and plastic sheeting;*
- *(m)* *sockets, socket covers and filling.*

Box 5 Changes to item coverage for terminals

406 • Untensioned Corrugated Beam Safety Fence (UCB)

The Sixth Edition Clause 405 is completely rewritten. The requirement for the longitudinal clearance to be removed as in Clause 405 is introduced and the requirement in respect of posts and assembly are omitted.

407 • Open Box Beam Safety Fence (OBB)

The Sixth Edition Clause 404 is completely rewritten. The requirements for fences adjacent to structures, adaptor pieces and assembly details are omitted.

408 • Double Rail Single Sided Open Box Beam Safety Fence (DROBB)

This type of safety fence was not covered in the Sixth Edition SHW although the Notes for Guidance did mention it.

Assembly requirements are similar to SHW 407 with the additional requirements of 408.4 relating to the relevant position of the upper and lower beam joint.

Definition

2 The term 'wire rope' shall mean the complete rope system for the wire rope safety fence comprising upper and lower ropes together with inherent component ropes of all types and tail ropes but excluding safety check ropes.

32 Itemisation

Group	*Feature*
I	*1 Wire rope.*
	2 Each type of post.
	3 Each type of intermediate anchorage.
	4 Each type of end anchorage.
	5 Each type of concrete foundation to post.
	6 Each type of socketed foundation to posts.

Box 6 MMHW Definition and Itemisation of wire rope (WR)

409 • Wire Rope Safety Fence (WR)

This type of safety fence was not covered in the Sixth Edition (see Boxes 1 and 6 for measurement aspects).

The ropes are positioned as set out in SHW Clause 409.2 with a maximum length of 627 m. Tensioning can only proceed when the Engineer is satisfied that the anchorage is secure and the temperature is within the range of 30°C to 10°C.

Rigging screws (with a minimum thread engagement of 25 mm) are adjusted to achieve the tension shown in Table 4/5 appropriate to the ambient temperature and agreed by the Engineer. The tension must be checked before bringing the fence into service and it must be retensioned if necessary.

***410* • Tensioned Rectangular Hollow Section Safety Fences (RHS)**

The requirements of SHW 410.1 are similar to the Sixth Edition Clauses 403.1, 403.2 and 403.4. The steel is to comply with SHW Clause 402.2.

Additional requirements cover tensioning. The Engineer has to be satisfied prior to commencement and the ambient temperatures must be between 10°C and 20°C. Tensioner assemblies must not be more than 70.5 m apart and each installation must incorporate at least one tensioner assembly. The tensioning procedure is set out in SHW 410(2) (iii)(a)–(h), but (d) introduces the impractical requirement that the anchorage bolts shall be tightened until the adjacent intermediate post 'just maintains its vertical position without signs of movement'.

411 • Concrete Safety Barriers

This type of barrier was not covered in the Sixth Edition (see Box 7 for measurement information). The clause covers British Concrete Safety Barrier (BCB), Permanent Vertical Concrete Safety Barrier (VCB) and Temporary Vertical Concrete Safety Barrier (TVCB).

Temporary Vertical Concrete Safety Barriers are shown in HCD and are to be used during construction where the road is subject to a speed limit of less than 80 km/h. Such use should be agreed with the Overseeing Department, and Appendix 4/1 should specify whether they are to be provided by the Contractor and retained by him on completion, provided by the Contractor but become the property of the Overseeing Department on completion or are provided by the Overseeing Department for the Contractor's use during the Works.

Group	*Feature*
I	*1 Each type of barrier.*
	2 Each type of termination.
	3 Each type of transition.
II	*1 Straight or curved exceeding 50 metres radius.*
	2 Curved not exceeding 50 metres radius.

Notes for Guidance — Safety Fences

The MMHW provides for three categories of curvature for payment purposes. Curves which are made up from infividual straight lengths of beams should not be considered to be small lengths of straight fence. They should be measured as curved fences within the Group II Features in MMHW. The radius is to be considered to be the radius equal to that of the arc which passes through the posts.

Box 7 Measurement — Concete Safety Barriers

Box 8 MMHW and Notes for Guidance

Notes for Guidance — Pedestrian Guardrails and Handrails

Curves which are made up from individual straight lengths should not be considered as curved elements but as straight guardrails or handrails.

Where the rails are actually curved they should be measured as curved guardrails or handrails as described by the specific radius.

Group III Feature 1 — Additional Information

Elements curved in plan to different radii.

412 • *Pedestrian Guardrails*

This clause was previously in the Sixth Edition 300 Series (see Box 8 for measurement details). Location and type are to be described in Appendix 4/2.

Table NG 0/2 List of Sub-clauses which permit Contract-specific requirements to be included in the Contract instead of the national ones stated, e.g. Sub-clauses state '. . . unless otherwise described in Appendix -/-'

403.9	Timing of submission of evidence of capability of anchorages in drilled holes.
404.4	(i) Timing of post foundation tests.

Table NG 0/3 List of Sub-clauses which require the Contractor to submit information to the Engineer.

Note. Information that the Contractor may submit when seeking the Engineer's approval is not listed in this table.

402.5	(v) Destructive testing of welds — provide copies of reports for earlier contracts.
402.7	(iii) Wire rope terminals — provide evidence of tensile tests.
403.9	Anchorages in drilled holes — provide evidence of capability.
404.4	(i) Post foundations tests — submit test results.

APPENDIX A Quality Management Schemes

2 Description: Supply and Erection of Fences (QGN 5000/5, formerly QAS 4630/5)

Certification Body: BSI Quality Assurance
PO Box 375
Milton Keynes
MK14 6LL

Specification: General fencing shall be in accordance with the 300 Series of the Specification and the following British Standard:

BS 1722: Specification for fences Parts 1 to 13.

Safety fencing shall be in accordance with the 400 Series of the Specification and the Drawings referred to in the Contract.

Note: This Scheme applies for general fencing and safety fencing. The part of this Scheme which relates to safety fences requires training on an approved course run through the Local Government Management Board or the Construction Industry Training Board.

3 Description: Manufacture of Safety Fencing Components (QAS 4630/204, QSS 32042)

Certification Bodies: BSI Quality Assurance
PO Box 375
Milton Keynes
MK14 6LL

Lloyd's Register Quality Assurance Ltd
Norfolk House
Wellesley Road
Croydon
CR9 2DT

Specification: Safety fencing components shall be in accordance with the 400 Series of the Specification and the Drawings referred to in the Contract.

APPENDIX E Departmental Type Approval/Registration

Approval Body: The Overseeing Department

Types of work, goods or materials for which proprietary products are required to have Departmental type approval/registration are as follows:

Description	**Code**	**Specification Clause**
Safety fences and barriers	A	401

Code Key:
A = Departmental type approval
R = Departmental registration

500 SERIES Drainage and Service Ducts

Introduction

One of the welcome changes in the Seventh Edition has been to accept that billing for drainage with all the adjustment items is a lengthy process. In order to simplify this the Notes for Guidance MMHW suggest that a tabulated format may be used. This can also be extended to manholes and chambers and an example of this table is included in the Fourth Edition of MMHW Notes for Guidance (see Table 1 on next page).

It is also recognised that the structural drainage and service ducts can be quite extensive. In order to clarify this the NG MMHW advise that the quantities should be scheduled either on a drawing or in an appendix and the interface with non-structural drainage clearly shown on the drawings. There is provision in Appendix 5/2 for the inclusion of a structural drainage/duct schedule.

With regard to alternative pavement types the NG MMHW confirm that it is not necessary to provide separate drainage bills for each permitted option. The measurement of drainage must be based on the thinnest pavement construction permitted by the options, irrespective of the actual construction.

Other significant changes to the 500 Series are that testing requirements now have to be described in Appendix 1/5. This is in line with the Seventh Edition overall philosophy. In addition to this there are two extensively amended Clauses 514 and 515 relating to Fin Drains and Narrow Filter Drains. These amended clauses contain details of the geotextiles permitted, the nature of composite drains, joints, pipes, backfill, dimensions, installation and handling, identification procedures, testing and finally the certification process.

The peculiar double cross reference for measurement of Filter Drains has now been amended in the MMHW Series 500, para. 14 (see Box 1).

Drains, Sewers, Pipes Culverts and Service Ducts

Measurement

14 The measurement of filter drains, excluding narrow filter drains, shall be the summation of their individual lengths measured along the centre lines of the pipe (or trench where no pipe is provided), between any of the following:

Box 1 MMHW Series 500

The Contractor has been given design responsibility for box culverts of spans up to 8m. This is in line with the avoidance of barriers to trade as set out in SHW Clause 106. This clause requires the structures to be listed in Appendix 1/10 with the drawings showing the designated outline of the structure.

Drains or culverts that are to be Contractor-designed are to be measured in accordance with Series 2500.

Flat invert corrugated UPVC pipes have been deleted, but these products possess a BBA certificate and therefore may be used (see SHW 501.7), and combined drainage and kerb systems have been modified and are now Contractor-designed and as such are to be listed in Appendix 1/11 with design parameters detailed in Appendix 5/5. Technical requirements for corrugated steel pipes are included in SHW Clause 501.4.

Table 5/1, listing the pipes and fittings, has been augmented to include pipes or fittings that have an Agrément Board Certificate and at the completion of the whole of the drainage works the Contractor has to supply to the Engineer schedules showing all the details of the pipes used.

The reference to provisional items to be included in the Bill of Quantities (Sixth Edition NG 502.2) for soft spots has not been included in this edition of the Notes for Guidance, but there is still a measured item for soft spots in the MMHW.

In the following notes on individual clauses, the use of italics in headings, text and clause number denotes areas of change.

Table 1 Extract from NG MMHW Series 500: Tabulated Drainage and Chamber Example

1. Drains and Sewers

Item	Description				Unit	Quantity	Rate	£	p
	'A' mm internal diameter drain or sewer specified design group 'B' in trench depth to invert exceeding 2 metres, but not exceeding 4 metres, average depth to invert 'C' metres Adjustment on this item for variation greater than 150 mm above or below the average depth of 'C' metres per 25 mm of variation in excess of 150 mm. Rate per metre 'D' (not to be extended)								
	'A' dia	'B' design group	'C' ave. depth	'D' adjust. rate					
21	150	6	2.625		m	54			
22	225	7	2.950		m	18			
29	300	7	2.875		m	78			
24	450	8	3.275		m	157			

2. Chambers

Item	Description						Unit	Quantity	Rate	£	p
	Chamber specified design group 'A' sub type 'B' with 'C' 'D' and frame depth to invert exceeding 'E' metres but not exceeding 'F' metres.										
	'A' design group	'B' sub-type	'C' cover grade	'D' cover type	'E' depth min.	'F' range max.					
76	2	—	grade A	cover	1	2	no	10			
77	3	a	grade A	cover	1	2	no	60			
78	3	b	grade A	cover	1	2	no	70			
79	3	c	grade A	cover	2	3	no	55			

501 • Pipes for Drainage and for Service Ducts

General

Box and piped culverts can be Contractor-designed and must not be specified where either a concrete or corrugated steel culvert would be technically feasible. The structure must be referenced in Appendix 1/10 and designed in accordance with SHW Clause 106 (see notes on 100 Series).

Drains constructed using corrugated steel tubes exceeding 900 mm dia. are excluded from this series and must comply with the 2500 Series dealing with Special Structures.

NG 501.2 advises that it is the duty of the Engineer to see that the requirements of the 500 Series are not inconsistent with any specifics included in a partially compiled AIP (Approval in Principle) form.

For plastic pipes the Engineer must specify the pipe stiffness and impact resistance in Appendix 5/1.

There have been some changes to the MMHW. There are four new items added to the item coverage. Three of these were added by the October 1991 Addendum to the Third Edition of the MMHW (see Box 2).

The item for supplying test certificates to the Engineer has been deleted and this is relied upon to be provided by the item coverage in the Preambles to the Bill of Quantities. In addition to this, item (e) has been amended to include rocker pipes.

The item coverage for the 1900 Series is shown in Box 3.

Drains, Sewers, Pipe Culverts and Service Ducts — Para. 12 —

Item coverage

(q) *pipe schedules;*
(r) *lubricants, packing, grouting and caulking;*
(s) *surveys and recordings;*
(t) *protective system (as Series 1900 paragraph 4).*

Box 2 MMHW Series 500

Protective System 4

Item coverage

(a) despatching paint samples to testing authority;
(b) shop procedural trials;
(c) site procedural trials;
(d) masking and other measures to protect adjacent untreated work and the removal of masking and other measures upon completion;
(e) joint fillers and treatment of joints;
(f) preparing materials for application;
(g) preparation of surfaces and coating at the place of fabrication and on Site;
(h) complying with any special requirements in respect of ambient conditions including the containment of dust and debris and for intervals between successive operations and applications;
(i) stripe coats;
(j) obtaining the correct dry film thickness of paint or other coating;
(k) preparation and supply of system and data sheets;
(l) facilities and assistance for Engineer's inspection.

Box 3 Extract from MMHW 1900 Series

The Contractor is to be responsible for ensuring that plastic pipes are not subject to deterioration due to sunlight between manufacture and installation in the ground. It could be difficult for the Contractor to guarantee that the pipes have not been so exposed during storage at the manufacturer's or supplier's yards.

Pipes for Drainage

A wide range of materials for pipes for drainage will normally be offered and the full selection will be detailed in Table 5/1 together with a schedule of the permitted pipe and bedding combinations from Appendix 5/1. *Pipes not included in Table 5/1 but with British Board of Agrément Roads and Bridges Certificate may be used as alternatives. The procedure for approving BBA certified pipes is not detailed in the SHW.*

The Contractor must supply the Engineer with a schedule showing details of pipes, joints and quality on completion of the whole of the drainage works. An appropriate item coverage is now included in the MMHW Series 500 para. 12(q) (see Box 2).

The Engineer must provide the basis of the hydraulic design on which the Contractor is to submit his proposals on pipe types. This must be detailed in Appendix 5/1.

For protection against acidic or sulphate attack, the design should consider limiting certain pipe materials or specifying sulphate resisting cement or bitumen coating. Other measures include protection for the invert of corrugated steel piped culverts, where stones are likely to be carried in the flow, by means of asphalt or concrete coatings. Protection for iron pipes in known acidic conditions can be provided by polythene sleeves.

Pipe runs of more than one type may exceptionally be used but the Engineer must consider the compatibility of the joints and the smooth inner integrity for rodding purposes.

Corrugated Steel Pipes Not Exceeding 900 mm Diameter
Corrugated pipes shall be manufactured from either bolted segments or *galvanised steel sheets* and if extra protection is required this must be detailed in Appendix 5/1.

The Engineer is responsible for specifying the metal thickness for corrugated steel pipes not exceeding 900 mm diameter of the lock seam fabrication type. This shall be 1.25 mm for galvanized steel but for the bolted segmental plate types SHW 501.4(i) states the thickness to be detailed in Appendix 5/1 and NG 501.4 advises that tables issued by manufacturers recommend thicknesses for various diameters and fill above the pipe. In view of the avoidance to barriers to trade as set out in Clause 106, it could prove difficult for the Engineer to comply with this.

With bolted segmental plates, SHW 501.6 requires that tightening may need to be repeated to achieve the necessary torque.

Pipes for Service Ducts
Pipes for service ducts shall be tested in accordance with SHW 509.10 *but other pipes can be used as an alternative if they possess the British Board of Agrément Roads and Bridges Certificate. The procedure for approval of BBA certified pipes/ducts is not clear.*

Ducts shall be fitted with draw rope of 5 kN breaking load and a 20 year life. Certain pipe duct materials have been excluded from the specification because of the difficulty of drawing cables through them.

Precast combined drainage and kerb blocks are specified in a new Clause 516 and these shall be Contractor-designed with the details listed in Appendix 1/11.

502 • Excavation for Pipes and Chambers

Where it would be unwise to disturb the ground or under heavy trafficked roads consideration must be given to thrustboring or jacking.

The itemisation and item coverage for Extra Over Excavation in hard material has been amended. This amendment was introduced in the October 1991 Addendum and serves to specify that the extra over is related to hard material in drainage as opposed to the main construction as the Sixth Edition SHW (see Box 4).

An issue which has caused some confusion, but which remains unchanged from the Third Edition MMHW, has been the adjustment item associated with the measurement of depths to invert, Group III Features. The MMHW does not make it clear whether the adjustment can result in a positive or negative value.

It is recognised that actual depths to invert of drains, sewers, piped culverts, service ducts and filter drains might vary from the averages described in the Bill of Quantities. Adjustment items are provided in the Bill of Quantities to evaluate the variances in depths to inverts between

Excavation in Hard Material

Group	Feature	
I	1	Extra-over excavation for excavation in Hard Material in drainage

Extra-over Excavation for Excavation in Hard Material

69 The items for extra excavation in Hard Material in accordance with the Preambles to Bill of Quantities General Directions include for:

Item coverage

(a) excavation in Hard Material (as Series 600 paragraph 23)

Box 4 Extra-Over Excavation in Hard material — Itemisation

the averages described and the actual depths calculated from site measurement. Adjustments may increase or decrease the amounts in the Contract Price.

The Sixth Edition NG 502.2 advised provisional items to be provided in the Bill of Quantities if soft spots were anticipated. This advice is not offered in the Seventh Edition NG. There is still provision to measure soft spots in the MMHW.

Backfilling of over-excavation is to be with only *ST1* mix concrete where compaction of alternatives is impractical.

The Engineer's agreement to batter slopes for excavation should not be withheld unnecessarily.

503 • Bedding, Laying and Surrounding of Pipes

Pipes have to be laid to the levels and gradients as shown on the drawings and schedules. Deviation from the line specified shall not be more than 20 mm at any point. This is an additional specification from the Sixth Edition SHW. Pipes shall not be laid on setting blocks unless on a concrete bed or cradle. Pipes and fittings are to be examined for damage before laying, with measures taken to prevent soil or other material from entering the pipes. Each pipe is to be anchored to prevent movement before the work is complete.

Corrugated coilable perforated pipes are only to be laid by a single pass drain laying machine. This will considerably restrict the use of this type of pipe.

A wide grading envelope is permitted for pipe bedding material, however to make savings in coarser materials a sand bed may be used. Distinctions should be made between bedding, haunching and surrounding and the requirements for backfilling.

Concrete surrounds to pipes should only be used exceptionally, essentially for protection only. *Pipe and bedding combinations shall be selected from Appendix 5/1. Bedding, haunching and surround material shall be as shown on HCD F1 and F2.* For pipes designated as Types B, F and S, the granular material shall comply with the gradings in Table 5/3 and have a water-soluble sulphate content of 1.9 g/litre.

Unless stated otherwise, bedding, haunching and surrounding of filter drains is to comply with the appropriate clauses, i.e. 503.3(i) and 503.3(iv); *however, granular materials used for bedding, haunching and surround shall have a 10% fines value of not less than 50 kN when tested in accordance with BS 812.*

Duct construction shall comply with details required in Appendix 5/2.

504 • Jointing of Pipes

Requirements for watertight or partly watertight joints are to be set out in Appendix 5/1 but surface water drain joints do not always need to be watertight. The criterion on surface water pipes is ingress of fine particles or root penetration. *Ducts which are required to be waterproof should be detailed in Appendix 5/2.*

The maximum length between flexible joints may have to be limited if movement is expected and the limits must be detailed in Appendix 5/1. As culverts need not necessarily be watertight, then details must be specified in Appendix 5/1 if this is now required. *If this is also the case with a Contractor-designed culvert then the Engineer would have to specify this in Appendix 1/10.*

Rigid joints require Engineer's approval but spigot and socket joints may be caulked with tarred rope. The sockets must be filled with mortar. Iron pipes can be caulked with lead wool.

If preformed filler is required in a concrete bed or surround it should comply with SHW 1015.

505 • Backfilling of Trenches and Filter Drains

The Notes for Guidance recommend that a wide range of materials should be specified so that local sources can be used as far as possible. Filter drains shall be backfilled as described in Appendix 5/1. Type A for subsoil drainage, Type B for intercepting surface water and Type C when a particular grading is specified.

Some methods of preventing stone scatter are shown on HCD B15. If adopted, these would require appropriate item coverages. If other than Class 1, 2 or 3 general fill material is required then this will need to be detailed in Appendix 5/1. If permeability tests are to be carried out, the requirements should be described in Appendix 5/1.

Compaction compliance is to be as SHW 612 with filter material compacted in 225 mm layers in accordance with Method 3 Table 6/4. Backfill material must not be heaped in the trench before being spread and power rammers are restricted in their use unless at a distance greater than 300 mm from any part of the pipe or joint.

Except in carriageways and paved areas topsoil thickness at the top of backfill must be stated in Appendix 6/8 if it is required to be different from the surrounding ground. For carriageways or paved areas, backfill is brought up to formation level or to a level as stated in Appendix 5/1.

The MMHW reference to Filter Drains has been rewritten and now has

Filter Drains

Measurement

14 *The measurement of filter drains, excluding narrow filter drains, shall be the summation of their individual lengths measured along the centre lines of the pipe (or trench where no pipe is provided), between any of the following:*

(a) the internal faces of chambers;
(b) the external faces of headwalls;
(c) the intersection of centre lines at junctions;
(d) the centre of gully gratings (or where no grating is provided the centre of the gully);
(e) the position of terminations shown in the Contract;
(f) the point of change of stage depth.

15 *The depth of filter drains shall be the vertical measurement between the invert (or the centre line of the trench bottom where no pipe is provided) and the following:*

(a) where the invert is below the Existing Ground Level — the Existing Ground Level or the Earthworks Outline whichever is the lower, except that where the finished level of the filter material is above the Existing Ground Level the measurement shall be taken to the finished level of the filter material;

Box 5 Filter Drain Measurement

Itemisation

Group	Feature
I	1 Filter drains.
	2 Filter material contiguous with filter drains.
	3 *Sub-base material.*
	4 *Lightweight aggregate infill.*
II	1 Different internal diameters.
	2 Different types of filter material.
	3 *Different types of sub-base material.*
	4 *Different types of lightweight aggregate infill.*

Box 6 Filter Drain Itemisation

its own measurement paragraph (see Box 5). Some of these amendments were introduced in part in the October 1991 addendum to the Third Edition MMHW and now have new paragraph references in the Fourth Edition. The significant change is paragraph 15(a). The Sixth Edition did not provide adequate measurement to Filter Drains in shallow embankments.

An amendment introduced by the Addendum and carried through with the Fourth Edition MMHW is the inclusion of Sub-base and Lightweight materials in the Groupings for Filter Drains (see Box 6).

506 • Connecting to Existing Drains, Chambers and Channels

Requirements for any special connecting pipes should be described in Appendix 5/1 and existing drains which are no longer required are to be grouted and sealed as directed by the Engineer.

Contractors are required to give notice to the Authority concerned before entering an existing sewer.

507 • Chambers

All chamber types are to be specified in Appendix 5/1 and constructed as shown in HCD. Testing of foul chambers for watertightness should be described in Appendix 5/1 and scheduled in Appendix 1/5. *All covers and gratings shall be described in Appendix 5/1 and comply with BS 497. If the Engineer requires these to be coated then the requirement must also be stated in Appendix 5/1. Brickwork for chambers has to be in accordance with SHW 2406, but if a different brick is required then it must be specified in Appendix 24/1.* The specifying of a different type of brick may give rise to a barrier to trade.

Precast chambers do not require strengthening by concrete surround but access shafts should be. If space is insufficient to allow compaction with acceptable material then concrete should be used.

If necessary the Engineer may vary the lengths of articulated pipe sections to satisfy SHW 507.15 and Table 5/6. *If adjusted covers or frames are required to be set at any other level than the new surface then this must be detailed in Appendix 5/1.*

The Sixth Edition SHW Clause 507.17, requiring notice to be given to Authorities before commencing work on Statutory Undertaker's chambers, has been deleted. Presumably the reference in Clause 506.2 is sufficient.

Chambers and Gullies

Group		Feature
I	1	Chambers.
	2	Gullies.
II	1	Specified design groups.
	2	Particular designs stated in the Contract.
III	1	*Depths not exceeding 1 metre.*
	2	*Depths exceeding 1 metre but not exceeding 2 metres and so on in steps of 1 metre.*
IV	1	Different types of covers or gratings.

Item coverage

(o) filling;
(p) notices;
(q) sealants (as Series 2300 paragraph 10);
(r) brickwork (as Series 2400 paragraph 4).

Box 7 Changed itemisation and item coverage

The MMHW itemisation has been amended, partly by the October 1991 Addendum, to reflect the fact that the depths for chambers can be to the bottom of the excavation. The item coverages have also been amended (see Box 7).

508 • Gullies and Pipe Junctions

No specific approval from the Engineer is required to install trapped gullies as was required in the Sixth Edition SHW. These can be allowed and should be described in Appendix 5/1 and be in accordance with HCD F13 and F14.

Gratings should be flat *and iron gratings are no longer required to be supplied uncoated.* The slots are not to be parallel to traffic. Brickwork to seat gully frames is to be properly constructed and in accordance with SHW 507.3.

With in-situ trapped gullies where permanent plastic shutters are used, the NG 508.2 requires the Engineer to ensure the trap is equal in all respects to that of a precast concrete or clay gully. Where in-situ gullies are formed with permanent shutters these must have a British Board of Agrément Roads and Bridges Certificate. Backfilling is to be as described in Clause 508.6 as Class 1. *The Sixth Edition SHW allowed Class 2 and 3.* Gully connection pipes of either rigid or flexible type shall not exceed 0.7 m in length with flexible joints for a distance of 2 m from the gully.

SHW 508.8 is a new Sub-clause which details adjustments or replacements of frames and gratings as Sub-clauses 4 and 5 to 6 mm below the adjoining surface, unless Appendix 5/1 details otherwise. The finished mortar bed shall be 10–25 mm thick with any final adjustment made by modifying the brickwork or using a different frame depth. The MMHW has been amended to include brickwork (see Box 8).

Gullies

Item coverage

(i) brickwork (as Series 2400 paragraph 4).

Box 8 Gullies — Item coverage

509 • Testing and Cleaning

The most significant feature of Clause 509 is that testing has to be '... as described in Appendix 1/5 or as required by the Engineer ...'. The latter part of this sentence has been amended from 'as directed by the Engineer' which was the case in the Sixth Edition. Clause 36(3) of the ICE Fifth Edition will be amended to reinforce the need for details of testing to be sufficient for tendering. The amended Clause 36(3) is shown in Box 9.

For testing, Clause 105 of the SHW and NG 105 detail the necessary requirements. NG 105.2 recommends that the tests required be abstracted from Table NG 1/1 and then scheduled in Appendix 1/5. Testing so described in the Contract will form part of the General Obligations as referred to in the Preambles to the Bill of Quantities (see Box 10).

Clause 36

Sub-clause (3) is deleted and replaced by the following sub-clause:

Tests

(3) Each test of materials or workmanship which is specified in the Contract as to be carried out by the Contractor shall be carried out by him at his own cost provided that the test in question is particularised in the Specification in sufficient detail to enable the Contractor to have priced or allowed for the same in his Tender. The cost of any test carried out which is:

(a) not so particularised in the Specification; or
(b) specified in the Contract and carried out by the Engineer; or
(c) not specified in the Contract;

shall be borne by the Contractor if the test shows the workmanship or materials not to be in accordance with the provisions of the Contract or the Engineer's instructions, but otherwise by the Employer.

Box 9 Amended ICE Fifth Edition Clause 36

Box 10 Preambles to Bill of Quantities para. 2 (vii)

General Directions

(vii) General obligations, liabilities and risks involved in the execution of the Works set forth or reasonably implied in the documents on which the tender is based.

Box 11 Item coverage changes

Drains, Sewers, Pipes Culverts and Service Ducts

Item coverage

(n) checking and cleaning;
(o) recording, staking and labelling;
(p) in the case of ducts fixing draw ropes, removable stoppers, marker blocks and posts;
(q) pipe schedules;
(r) lubricants, packing, grouting and caulking;
(s) surveys and recordings;
(t) protective system (as Series 1900 paragraph 4).

Testing of pipelines is done by air tests and if they are rejected due to a failed air test the Engineer may agree to a water test as an acceptable alternative. If accepted after testing, what effect does the SHW 509.5 video survey have? One of the ambiguities created by the Sixth Edition SHW was the fact that the video survey was required on completion of the whole of the works. It could be argued that Clause 36(3) of the ICE Fifth Edition was not applicable. In addition there were no appropriate item coverages for this work. *The MMHW has been amended to include an item for this (see Box 11).* There is no guidance as to how the video survey is assessed and how any defects will be paid for.

With regard to a mandrel check, SHW 509.4, the item coverage in the MMHW mentions only checking and cleaning. Does this checking refer to the mandrel test? In view of NG 105.3 the Contractor could claim payment for air tests and mandrel tests if the details are not included in the new Appendices 1/5 and 1/6. NG 509.3 suggests that the Engineer is advised to liaise with the Drainage Authority to ensure acceptability.

Pipes and filter material are to be left clean and free from silt 'at all times'. Tests for partly watertight joints should be carried out before the pipe is laid in order to measure the water escaping. This test also has to be detailed in Appendix 1/5.

The calculation for the acceptable seepage of water from a partly watertight joint as described in the Sixth Edition SHW Clause 509.7 has now been corrected.

510 • *Surface Water Channels and Drainage Channel Blocks*

Details of the design of drainage channel blocks shall be given in Appendix 5/1 and must also comply with Clauses 1103 and 1101 of the SHW. Comments on these will be addressed in Series 1100.

511 • Land Drains

The preferable arrangement for carrying out work on disturbed existing land drains adjoining the Works is to have these drains separate from the new road drainage system. The reinstatement is then a matter for the District Valuer. It may, however, be cost effective to link the existing drains into the new drainage system.

Existing land drains which have been severed as a result of the works shall be connected into the new drain as described in Appendix 5/1. Pipes disturbed by the works shall be re-laid. How is this paid for? The MMHW has no item coverage for relaying. Disused ends shall be sealed with concrete.

Work carried on outside the works as defined by the Conditions of Contract would need to be detailed in Appendix 1/7.

512 • Backfilling to Pipe Bays and Verges on Bridges

Any special requirement should be specified by details on drawings and referenced in Appendix 5/1. Compaction and laying are to be in accordance with SHW 505.4 and 505.5.

513 • Permeable Backing to Earth Retaining Structures

Unless detailed otherwise in Appendix 5/1, backing can be either 300 mm thick granular Type A, Type C as SHW 505, porous no-fines in-situ concrete 225 mm thick as SHW 2603 or precast concrete blocks to BS 6073. If the filling adjacent to the structure is cohesive Class 7B, or Class 7B, or chalk, the backing shall be 300 mm of fine aggregate grading C or M. *Type C granular backing has now been added as a design initiative so that the Engineer can design a filter compatible with the adjacent material. Fin drains are not allowed as backing as their design life has not yet been established.*

514 • Fin Drains

Fin Drains and Narrow Fin Drains

(d) protection from ultraviolet light;

Box 12 Item coverage para. 23(d)

The Contractor is to be allowed a choice as detailed in HCD and must comply with the requirements of Appendix 5/4. If the Engineer requires to qualify this then exclusions must be stated in Appendix 5/4.

Drain materials are to be protected from short-term exposure to ultraviolet light. The criterion for this is referenced in SHW 514.11 as a cumulative value of 50 hours. The MMHW has an item coverage for this (see Box 12).

Labelling the side of water entry and direction of flow may be required. The pore size of the geotextile selected filtration criteria is to be compatible with the adjacent soil or construction layer, but with very fine grained soils, the criteria in NG 514.2 will result in such small pores that sufficient permeability may not be achieved. The BS test to determine pore size may be inappropriate for some textiles and if this is the case then wet sieving techniques may be used for determination.

Permeability requirements of the geotextile should incorporate a margin of safety to allow for clogging.

The design flow capacity of composite drains should allow for infiltration through the pavement and other nearby sources but as infiltration rates through pavements are yet to be verified, a value of not less than 0.2 times the mean intensity of a one-year, two-hour rainfaill should be assumed.

Fin drains are normally at constant depth and gradient to follow the road. They should be capable of being joined both longitudinally and laterally into pipe systems and where edges may become exposed they shall be protected with geotextile wrapping. For drains installed in a trench, the backfill material shall be the as-dug material. If, however, the permeability of this as-dug material when compacted would render the drain inefficient or contain stones larger than 100 mm then an alternative material should be used.

These drains are not to be used during construction for disposal of run-off water.

Fin drains must be the subject of a British Board of Agrément Roads and Bridges Certificate and during construction assembly areas are to be kept clean and material which becomes contaminated is to be replaced. No criteria for contamination are given so this must presumably be at the decision of the Engineer.

After installation the drain must be marked with marker tape in positions shown on the HCD. The Contractor is required to supply to the Engineer consignment details for each delivery to site of fin drains. The Sixth Edition

Box 13 Fin Drains and Narrow Filter Drains — Itemisation and item coverage

Fin Drains and Narrow Filter Drains

Group	Feature
I	1 Fin drains. *2 Narrow filter drains.*
II	1 Specified design group. 2 Particular designs stated in the Contract.
III	1 Depth not exceeding 1.5 metres.

Item coverage

(a) geotextiles and cores;
(b) backfilling and compaction;
(c) filter drains (as this Series paragraph 18);
(d) protection from ultraviolet light;
(e) marker tapes;
(f) lapping and jointing;
(g) connections, attachments and fittings;
(h) treatment at chambers, gullies, pipelines and the like.

MMHW itemisation for fin drains has been amended accordingly, some of the amendments were introduced in the Addendum and now extended in the Fourth Edition (see Box 13).

The tests required for fin drains are for:

(i) Contact area of the drainage core.
(ii) Determining the thickness of the drain under specified normal and shear stresses.

Variations in the tests can be made with prior consultation to the manufacturer by the British Board of Agrément. Under what condition would this be acceptable and what is the position of the Contractor? NG 514.7 states that little experience is available in interpreting test results and that this should be taken into account in analyzing results. These tests should be detailed in Appendix 1/5 although the SHW makes no reference to this. A note to compilers in Appendix 1/5 comments that not all tests would be included in NG Table 1/1, routine tests carried out by manufacturers and suppliers in compliance with a BS or other standard or specification would not be listed.

The Contractor should supply copies of the BBA certificates to the Engineer so that compliance with the Contract is achieved.

515 • *Narrow Filter Drains*

The Contractor is to be allowed a choice as with fin drains with limitations detailed in Appendix 5/4. Short-term exposure to UV light is not allowed but there are no criteria for 'short-term'. Does SHW 514.11 apply? Evidence is required by the Engineer to re-establish the integrity of the material after UV exposure but the nature of the evidence is not specified. Does this mean all the tests as in Clause 514 have to be carried out?

During installation, rocks or other hard protuberances in the bottom of the trench are to be removed, excess cut is to be backfilled with suitable or imported material. Marker tapes are to be installed as for fin drains.

Geotextiles used shall have a British Board of Agrément Roads and Bridges Certificate.

516 • *Combined Drainage and Kerb Systems*

The Contractor is to design these from the list in Appendix 1/11 as required by SHW 106 and also the design requirements in Appendix 5/5. The extent of the design must be clearly defined. A designated outline will not be required as combined drainage and kerb systems are classed as Structural Elements (see Clause 106 comments). This removes the complexities of designated outlines but leaves the problem of how to show these without involuntarily specifying a proprietary material.

Proprietary systems are allowed but with joining and laying in accordance with the manufacturer's instructions. On completion of the works the system is to be cleaned out with high pressure water jetting or other approved means.

Table NG 0/2 List of Sub-clauses which permit Contract-specific requirements to be included in the Contract instead of the national ones stated, e.g. Sub-clause '. . . unless otherwise described in Appendix -/-'

Drainage and service ducts	
501.2	Type of pipe between consecutive chambers.
501.4	Minimum thickness of corrugated steel pipes.
501.7	Internal diameter of pipes for service ducts.
503.3 (v)	Materials for bedding haunching and surrounding filter drains.
504.2	Watertight joints for ducts.
505.2	Backfilling of trenches.
505.6	Removal of supports to trenches.
507.4	Concrete for in-situ concrete chambers.
507.16	Setting of chamber frames and gratings or covers.
508.5	Gully grating profile.
508.8	Setting of gully frames and gratings or covers.
511.4	Filling to disturbed mole channels.
512.1	Backfilling to pipe bays and verges on bridges.
513.1 513.2	Permeable backing to earth retaining structures.
514.9	Backfill material for fin drains.
514.10	Dimensions and drain slope angle for fin drains.
515.6	Dimensions and drain slope angle for narrow filter drains.

Table NG 0/3 List of Sub-clauses which require the Contractor to submit information to the Engineer.

Note. Information that the Contractor may submit when seeking the Engineer's approval is not listed in this table.

Drainage and service ducts	
501.3	Pipes for drainage — provide a schedule.
506.1	Connecting to existing drains — record positions and hand copy to Engineer.
506.2	Connecting to existing drains — give notice to responsible authority.
509.5	Foul drains — supply recording of survey by video camera.
509.9	Concrete pipes exceeding 900 mm internal diameter — submit manufacturer's test certificates.
511.3	Land drains — give notice if blocked or defective.
514.12	Fin drains — supply consignment details.
515.8	Narrow filter drains — supply consignment details.
516.3	Combined drainage and kerb systems — provide evidence of suitability of application.

APPENDIX A
Quality Management Schemes

4 Description: Manufacture of Industrial Fasteners and Associated Items (QAS 3137/8)

Certification Body: BSI Quality Assurance
PO Box 375
Milton Keynes
MK14 6LL

Specification: Nuts, bolts and fixings shall be in accordance with the 300, 500, 1300, 1800, 2200 and 2500 Series of the Specification and shall be manufactured to the requirements of the following British Standards:

BS 3410 Specification for Metal Washers for General Engineering Purposes;
BS 3692 Specification for ISO Metric Precision Hexagon Bolts, Screws and Nuts:
BS 4190 ISO Metric Black Hexagon Bolts, Screws and Nuts;
BS 4320 Specification for Metal Washers for General Engineering Purposes:
BS 4395 Specification for High Strength Friction Grip Bolts and Associated Nuts and Washers for Structural Engineering;
BS 4933 Specification for ISO Metric Black Cup and Countersunk Head Bolts and Screws with Hexagon Nuts.

Alternatively nuts, bolts and fixings shall be manufactured to the dimensions and tolerances of the British Standards listed above, using materials to the following Standards:

BS 6105 Specification for Corrosion-resistant Stainless Steel Fasteners;
ASTM A325 Specification Highstrength Bolts for Structural Steel Joints.

Note: 1. Where the Contractor can demonstrate that the fastener required is made by less than three firms within this Scheme, the requirement to comply with the Scheme shall not apply.
2. Where the Contractor obtains fasteners from a stockist, the stockist shall be registered under Part 1 of the BSI Registered Stockist System. The System requirement shall be 'Level A, Quality Assured Material with Lot Traceability' (P00012). Where the Contractor can demonstrate that the fastener required is supplied by less than three stockists within this System, the requirement to comply with the System shall not apply.

APPENDIX B
Products Certification Schemes

BS Ref	Title
65	Specification for vitrified clay pipes, fittings and joints.
437	Specification for cast iron spigot and socket drain pipes and fittings.
497	Specification for manhole covers, road gully gratings and frames for drainage purposes.
124	Manhole steps. Part 1: Specification for galvanised ferrous or stainless steel manhole steps. Part 2: Specification for plastics encapsulated manhole steps.

APPENDIX C
British Board of Agrément Roads and Bridges Certificates

Description	SHW Clause
Pipes for drainage and/or service ducts other than those listed in Tables 5/1 and 5/2	501
Permanent shuttering for road gullies	508
Fin drains and constituent materials for edge of pavement drainage	514
Geotextiles for use in narrow filter drains for edge of pavement drainage	515

600 SERIES Earthworks

Introduction

As in the Sixth Edition, this 600 Series aims to provide the Contractor with sufficient information to enable the final product to be that required by the Department of Transport. This requires the contract documents to contain detailed information so that the Contractor is aware of which materials can be used, the limitations on their use and the likely availability of such material from within the scheme.

Whilst the principles are unchanged the designer now has available two new documents to assist him in the preparation of the contract documents and which fill a gap in the information available for the Sixth Edition. These are the 'Earthworks — Advice Note HA 44/91' which provides technical and background information and the Notes for Guidance to the Method of Measurement for Highway Works which provide useful advice on the principles of the earthworks measurement.

In designing the earthworks, consideration should be given to the adjustment of the precise alignment so as to achieve the best earthworks balance with the aim of minimising the import of materials by utilising almost all the materials arising from site, with treatment if necessary.

If imported material is required the specified characteristics should be adequate for performance but not unnecessarily restrictive and disposal off-site should be minimised by the use of surplus material in landscape areas (see Box 2).

The Engineer should match for use in fill areas materials likely to arise from excavation but should not normally indicate from where on site materials are to be obtained (Box 2).

The designer should specify, through Appendix 6/1, the characteristics of acceptable material so that the Contractor can determine the acceptability of a material by relatively simple and robust tests. Appendix 6/1 should list only those specific properties required by the Engineer to meet his design requirements omitting those which are unnecessary (see Boxes 3 and 5).

An example of a completed Appendix 6/1 is included in Annex A

Box 1

As with the Sixth Edition this specification requires the matter of testing to be clarified in Appendix 6/1, and HA 44/91 recommends that the Contractor should be made responsible for classification and determining acceptability using the designer's criteria set out in Appendix 6/1. It should be recognised, however, that the Contractor is still responsible under Clause 36(1) of the Conditions of Contract for ensuring that only acceptable materials are incorporated into the works even though the Engineer may be responsible for classification under Appendix 6/1.

So that the Contractor may properly allow for the costs incurred the designer must identify those tests which he is to carry out together with their frequency (see Box 4). This information must be specified clearly as payment depends upon whether each test is 'specified in the Contract as to be carried out by the Contractor' and 'is particularised in the Specification in sufficient detail to enable the Contractor to have priced or allowed for the same in his Tender' (ICE Clause 36 as modified by the MCD).

The DTp. suggest that the designer's estimate of the frequency of testing should be generous although this will marginally increase the tender values

3.35 Where there is an obvious excess of acceptable site material the Designer should make every effort to reduce the quantity of imported selected fills by means of designs using lesser quality but acceptable materials. An example would be to design a slightly more costly structure which could tolerate acceptable site arising material rather than expensive imported granular backfill material. Where selected materials arise on site the Designer should not indicate their location but he must be able to justify his design and may have to back up his decision in the event of a dispute. It can be seen that there is no necessity for the Designer to break down acceptable material into the various classes in the excavation but nevertheless he should be satisfied that the materials available meet the design requirements.

3.36 Sufficient detail must be given on the drawings for the Contractor to assess how much of each class of fill material is required and to determine where it is to be placed. The Designer will need to determine quantities of each class for his own purposes. The drawings will need to show where each class of material is to be placed, but the quantities shown will only be those required by the MMHW.

Box 2 HA 44/91 — Basic Design Parameters

4.2 Clause 601.1 SHW defines acceptable material as that which meets the requirements of Table 6/1 SHW and Appendix 6/1 SHW, and both Table and Appendix have been arranged so as to allow the Designer to best utilise the available materials on each individual scheme. Table 6/1 SHW divides both on-site and imported materials into 9 principal classes which are further sub-divided for compaction purposes or because of particular properties or applications. It also lists the criteria and relevant tests whereby a material may be identified as belonging to a particular class, together with the limiting values for some of those criteria which define whether the material is acceptable or not. Appendix 6/1 SHW is the means whereby the Designer can state values for those acceptability limits not already given in Table 6/1 SHW for the remaining criteria tests. This Appendix gives the Designer flexibility by allowing him to vary the acceptability limits of any material that would be acceptable in some circumstances and locations and not in others which may prevail on a site. Figure 4/1 sets out a recommended sequence for the use of Table 6/1 SHW and Appendix 6/1 SHW in the classification of materials and the setting of the acceptability limits for each property mentioned.

4.32 When preparing Appendix 6/1 for the scheme, the Designer should use the same format as Table 6/1 SHW, but only list the classes of material that he intends to use, and for these classes, only list those properties required for acceptability testing including those for which limits are already given in Table 6/1 SHW. Therefore Appendix 6/1 will give a complete list of acceptability criteria and limits for each material to be used in that scheme. If different or alternative materials are later specified the full details of properties should also be given in this form. Those criteria and material classes not required should not be included.

Box 3 HA 44/91 Part 1 — Format of Appendix 6/1

with little realistic prospect of recovery if the actual quantities are less.

As in the Sixth Edition, the 600 Series uses a mixture of 'end product' and 'method' specification to define what the Contractor must comply with, although the Engineer may well be checking that the result produced by the 'method' specification meets the design parameters adequately.

Generally there are surprisingly few changes to the 600 Series compared with the Sixth Edition although there is a general improvement in clarity on where information should be provided (appendices and drawings rather than 'as described in the Contract') and increased reference to British Standards in the testing clauses.

The processing of unacceptable material by the Contractor has been introduced (see Box 6), ground anchorages and crib walling are now to

3.6 If the Contractor is given the responsibility for determining the acceptability of all the materials used on site, he will need to carry out the required sampling and testing to prove to himself and to the Engineer that the correct materials are being used. Therefore the Engineer must specify what tests the Contractor shall carry out and give a frequency of testing for each test together with any other relevant information. Providing the contract makes it absolutely clear what and how much testing the Contractor is expected to carry out, then payment is allowed under Clause 36 of the CoC. For this reason sampling and testing has not been included in MMHW item coverage and no separate BoQ items are included for them. Suggested frequencies of testing are given in Paragraphs 3.8 to 3.12 but the Designer would be well advised to ensure that his estimate of the quantity of testing given in Appendix 6/1 is generous. Whatever sampling and testing is carried out by the Contractor, this does not affect the Engineer's own tests, although the frequency of sampling and testing by the Engineer's staff should not be as high as the Contractor's. Whoever is made responsible for the sampling and testing, all the relevant information should be included in Appendix 6/1.

3.8 If testing is the responsibility of the Engineer then the Contract should include details of all the apparatus and equipment required to equip the Engineer's site laboratory, as well as the usual services. If testing is the Contractor's responsibility, then in addition to the Engineer's laboratory, a list of tests together with the source reference for each test (BS, ASTM, SHW etc) which the Contractor is expected to carry out must be included in Appendix 6/1. Details about the submission of test results to the Engineer must also be set out. It is already the Department's policy that all permanent testing laboratories should be accredited by the National Measurement Accreditation Service (NAMAS) as part of their Quality Control policies: accreditation of site laboratories will follow at a later date. No mention should be made as to where the Contractor should carry out his testing ie his own site laboratory, independent testing laboratories etc, this should be left for him to organise.

3.9 In order for the Contractor to be able to include for testing in his rates, the frequency of each test should also be stated.

Box 4 Testing for Classification and Acceptability — HA 44/91

4.7 When establishing the limiting values of the acceptability criteria the Designer should consider a number of factors which could influence his choice of values and therefore his choice of fill:

i. Earthworks balance, haul distances and ease of transportation within the site.
ii. Availability of imported material from local borrow pits and ease of transportation to the site.
iii. Ability of the material to withstand site and construction traffic. Information regarding the effect of earthmoving plant on soil conditions can be found in TRRL LR 1034, SR 522 and RR 130.
iv. Ability of material to compact to a satisfactory density (grading and natural moisture content).
v. Moisture content and susceptibility to moisture.
vi. Frost susceptibility and rate of water supply through underlying soil.
vii. Possibility of improvement by groundwater lowering, drying.
viii. Chemical nature of material and effect on adjacent material, structures, pipes etc.
ix. Availability of material for landscaping environmental bunds etc.
x. Location of material within the fill area. It may be possible to use material in the core of an embankment.

Box 5 Assessing Acceptability Limits — HA 44/91

be designed by the Contractor and a new clause to cover determination of constrained soil modulus has been added.

Some changes have been made to Table 6/1: Classes 1B and 6F1 now exclude chalk; there are new Classes 6Q and 7H and additional properties

Processing of Unacceptable Material Class U1

24 The term 'processing' shall refer to treatment whereby material arising from the Site is rendered acceptable for a particular use in the Works by mechanical, chemical, hydraulic or other means.

25 The unit of measurement shall be:

(i) processing of unacceptable material Class U1 . . . cubic metre.

26 The processing of unacceptable material Class U1 shall be measured only when the Contract specifically requires particular material to be obtained for use in the Works by processing. Other processing carried out by the Contractor shall not be measured. The measurement of processing of unacceptable material Class U1 shall be the volume of the void required to be filled with the processed material.

Box 6 MMHW Processing of Unacceptable Material

in respect of 6H, 6I, 6J, 6K, 6L, 6M, 7C and 7D. The properties of fill for use with metal components in reinforced earth and anchored earth structures have been considerably changed in Table 6/3.

Relatively minor changes have been made to, amongst others, the clauses dealing with testing the density of compacted fill using nuclear gauges, compaction of stabilised layers in cutting, rates of seeding and earthworks to corrugated buried structures.

The specification deals very largely with what the Contractor can and cannot do and rarely strays into areas of measurement or payment (particular attention has been paid in the Seventh Edition to ensure that this is the case). These are the province of the Method of Measurement which sets out the rules for the compilation of the Bill of Quantities and the basis for payment.

Both documents are designed to complement each other and all activities that the Contractor is required to carry out are, in theory, covered by the Method of Measurement. It must be realised, however, that a change in the SHW will often require a change to the MMHW if the Contractor is to allow properly for that change in his prices.

Series 600 of the MMHW sets out precisely how the earthworks are to be measured and billed and what the Contractor must allow for in pricing a particular item. In measuring earthworks in particular it must be appreciated that the way in which the bill quantities are calculated will often not correspond with the actual quantities on site. This is a deliberate attempt to simplify and rationalise the measurement of the work and the real complexities will be reflected in the Contractor's rates rather than in the way the work is measured.

The basic principles of the MMHW are unchanged and the new document incorporates most, but not all, of the changes introduced in the Addendum to the Third Edition issued in October 1991. New items are introduced to cover the processing of acceptable material (see Box 7) and breaking up and perforating redundant pavement. The item coverage is extended in a number of cases and some of the previous pitfalls have been avoided by clearer wording of the imported fill and deposition items.

Surprisingly, the item for proof demonstration of capping introduced through the Addendum to the Third Edition has not been included and the clarification that this brought is lost. The item coverage for 'compaction' has been adjusted to compensate (see Box 29) and the Model Contract Documents will amend Clause 36(3) of the ICE Conditions of Contract to suit (see Box 31).

There is a new definition of 'hard material' (see Box 7) and this together with the explanation in the MMHW Notes for Guidance should help in this difficult area.

A common omission from Sixth Edition documents was the 'Earthworks

7 Preamble 13 to the Bill of Quantities sets out three methods of designating Hard Material for measurement purposes:

(a) designated strata
(b) designated deposits with limits shown on the Drawings
(c) existing pavements, footways, paved areas and foundations.

The selection of (a) or (b) above is achieved by applying professional judgment to borehole data and other sources of information to determine those identifiable strata and deposits which are likely to create significant costs relative to the excavation of other materials in the Works. It is intended that the results of this judgment should be included in the Contract. Admeasurement is based on that information and the Contractor is deemed to have made whatever allowances are regarded as being appropriate in his rates and prices.

The compiler should ensure that only one method of designation is used for any particular material. Once a strata or deposit has been designated as Hard Material it is not subject to reclassification. Conversely, the fact that a material similar to that designated as Hard Material in a deposit within defined limits shown on the Drawings, may be found elsewhere does not indicate that it will be measured as Hard Material in the other location.

Designation of material as Hard Material is for measurement purposes and is not intended to indicate that the material has any particular level of strength, bearing capacity of other characteristic.

Where Hard Material is designated by reference to named strata alone the total quantity excavated from within those strata is subject to admeasurement. Where deposits are designated by limits shown on the Drawings that volume is measured and paid for as Hard Material. For both methods of designation the material actually excavated may not fall within the definition of Hard Material as set out in sub-paragraph 1(i)(ii) of Chapter 1. Hard Material designated under Preamble 13(c) ie. existing pavements, footways, paved areas and foundations is subject to admeasurement but excluding any unbound materials within the pavement, footway, paved area of foundation.

Box 7 Hard Material — Changes to designation

Schedule', largely because it was included in the Library of Standard Item Descriptions. There should now be no doubt that the designer must provide this detailed information (but see Box 8) as it is specifically required in the Notes for Guidance (a completed schedule is included in Annex A).

As with the Sixth Edition, the designer has the responsibility of providing a great deal of information through the appendices and drawings. The Advice Note sets out some of these:

3.39 For a scheme which requires the excavation of cuttings and construction of embankments the following information will need to be shown

(i) Areas where topsoiling is required and the specified thickness and whether the slope is greater than 10° or less.
(ii) Areas as above for grass seeding, turfing or hydraulic mulch grass seeding.
(iii) Areas where any previous pavement is to be perforated or broken up and left in situ.
(iv) Areas where any previous road pavement is to be removed and to be measured as 'hard material'.
(v) Areas where any special treatment is required such as:
  — benching;
  — special embankment foundations — dig out and starter layers, backfill, selected fills, etc;
  — lengths of strengthened or reinforced embankment shoulders;
  — surcharge;
  — cut/fill transition zones.

Unfortunately the suggested method of checking that the schedule corresponds to the Bill of Quantity figures included in HA 44/91 para. 3.46 is seriously flawed and should only be used with extreme care. These errors have been drawn to the attention of the DTp. who are considering issuing an amendment in 1993.

Reference to the MMHW rules of measurement will show that the principles used in the check are incorrect and the difficulties are compounded by errors in the column numbers.

The calculation for:

F only works with a surplus of acceptable

D_E is wrong in principle as it requires that 'deposition' is measured on import.

C is also wrong in principle as 'compaction' must be measured irrespective of whether the particular material is to be compacted.

E+I is similarly wrong in principle as it assumes import is measured as deposition.

Column numbering of R24, S7 and S6 in various calculations is incorrect.

Box 8 Serious Errors in para. 3.46 HA 44/91

(vi) Earthworks limits and batter slopes required.
(vii) Landscape areas and extent of amenity bunds.
(viii) Noise measuring stations.
(ix) Extent of earthworks to structures.
(x) Extent of earthworks to side roads and interchanges.
(xi) Extent of earthworks in main carriageways.
(xii) Quantities of excavation, deposition and hard material, with a clear statement as to what is, or if necessary, what is not designated as hard material for payment purposes.
(xiii) Details of special treatments (see (v)).
(xiv) Exploratory holes in plan and section showing groundwater strikes, standpipe/observation well levels, extent of hard material, defined as either strata or within limits defined by levels, shear strength, MCV and Atterberg Limit test parameters.

Para 3.40 of the Advice Note then sets out the type and scale of the drawings where the above information should be shown.

In the following notes on individual clauses, the use of italics in headings, text and clause number denotes areas of change.

601 • Classification, Definitions and Uses of Earthwork Materials

Earthwork materials are classified as either acceptable or unacceptable. *A new Sub-clause introduces the processing of unacceptable material to render it acceptable (see Boxes 6 and 10 for new measurement rules).*

Table 6/1 is the key to classifying materials supplemented by Appendix 6/1 (see also Box 3). Acceptable materials are classified into nine principal classes:

> Class 1 and 2 — general fills — the greater part of materials normally encountered — incorporates chalk unless designated as Class 3 in Appendix 6/1. *Class 1B now excludes chalk.*
>
> Class 3 — chalk which is likely to be degraded by normal construction traffic.
>
> Classes 4 and 5 — landscaping and topsoiling.
>
> Classes 6 to 9 — selected fill for specific purposes. *6F1 now excludes chalk and there are new Classes 6Q and 7H with additional properties in respect of 6H, 6I, 6J, 6K, 6L, 6M, 7C and 7D.*

The Engineer may further sub-divide the classes if it is advantageous to do so.

5.58 PFA is an industrial by-product and for the purposes of SHW Table 6/1 is designated as Classes 2E and 7B material, or Classes 7G and 9C if cement stabilization is required (see Chapter 10). The sulphate content of PFA is likely to be such that SHW Clause 601 is relevant. The dimension restriction on depth below sub-formation and formation in SHW Clause 601 has been introduced for the following reasons:

i. because of the grain shape and size the upper layers of PFA are difficult to compact;
ii. freshly placed PFA behaves in a manner similar to a pure silt and, if not protected, may liquify following wet conditions;
iii. capping and sub-base materials tend to be relatively permeable and a layer of general fill over PFA is considered desirable to provide some protection.

5.59 The dimension may be marked on the drawings but it is recommended that SHW Appendix 6/3 is more appropriate. A figure of 600 mm is considered acceptable, and any reduction needs to be carefully considered.

5.61 SHW Clause 601 enables the Engineer to keep a record of the sources, and places the onus on the Contractor to provide the requisite data including the natural moisture content, and the optimum moisture content and maximum dry density of each consignment. Where field densities are being taken the top 100 mm of PFA should be removed before testing.

5.29 Minestone, if compacted to the SHW, will not spontaneously combust as oxygen will be excluded. Many examples exist of hot bunt minestone being satisfactorily compacted to form stable embankments. Minestone can therefore make very good all-weather general fill provided it is of sufficient quality. Some shales may be prone to softening and swelling if exposed to moisture and weathering when stockpiled. It is the Department's policy that the Designer should identify possible sources of this material for imported fill and ensure that Tenderers are aware of them.

Box 9 PFA and Minestone — HA 44/91 Chapter 5

36 The measurement of disposal of unacceptable material *Class U1* shall be the volume of unacceptable material *Class U1* excavated from within the Site and measured under this *Series less the volume of processed unacceptable material Class U1 calculated in accordance with paragraph 21 of this Series.*

37 *The measurement of disposal of unacceptable material Class U2 shall be the volume of unacceptable material Class U2 excavated from within the Site and measured under this Series.*

Box 10 MMHW — Changed Definitions

Unacceptable materials are either U1 or U2, the latter being hazardous material requiring special treatment. The details are to be set out in Appendix 6/2. Maximum particle size should be no more than two-thirds of compacted layer thickness with restriction on cobble size in verges.

There are new definitions of chalk and PFA and reference is to Appendix 6/1 rather than the Contract. There are also changes in test details for sulphates and in SO_3 limits where these are in proximity to metallic items.

Unburnt colliery shale is specifically allowed and excluded from the unacceptable definition of material susceptible to spontaneous combustion. *The requirements are listed in Appendix 6/1 (see Box 9).*

Pulverised fuel ash should not be placed in proximity to the sub-formation or formation. The distance is to be specified in Appendix 6/1 (see Box 9). The formation is the top surface of capping or earthworks unless shown otherwise *on the drawings.*

602 • General Requirements

Any special requirements on classification are to be shown in Appendix 6/1, e.g. who is responsible for classification — the Engineer or the Contractor (see Box 4).

Box 11 HA 44/91 Keeping earthworks free of water

3.38 In his prices, the Contractor is to include the requirements of SHW Clauses 602.15 and 16. However items need to be included in the BoQ for any requirements under SHW Clause 602.17. If the Contractor elects to render unacceptable material into acceptable (SHW Clause 602.7) then he should be paid as though the material concerned was disposed of and a similar quantity imported. If the Engineer wishes to carry out a similar operation then, under SHW Clause 602.17, the necessary requirements shall be set out in SHW Appendix 6/1 and paid for by an item in the BoQ.

The Contractor should only employ suitable plant and working methods to maintain the nature of acceptable material. If trial pitting is required this should be stated. As the item coverage for trial pits includes excavation in hard material, enough information should be provided to enable the tenderer to price this work.

Sub-clauses 3, 4, 5 and 7 have been reworded to cover processing of unacceptable material U1 where required by the designer (see Box 11 for meaurement when the Contractor elects to treat U1). Sub-Clause 5 excludes topsoil being run to spoil *although DTp. HQ may now give permission if requested. This will need to be considered and the SHW and MMHW amended as necessary*. Hazardous material is to be disposed of as described in Appendix 6/2.

The stability of excavations must not be affected by stockpiling, etc.

Topsoil is to be stripped to depths for 5A material as described in *Appendix 6/1*. If it is to be left in place this should be shown on the drawings *and detailed in Appendix 6/1. This will create the same difficulties as the Sixth Edition if the depths are not accurate. There is a change in the wording of the measurement paragraph relating to the excavation of topsoil.*

Topsoil is to be re-used immediately wherever practicable or stockpiled with specified restriction on height, trafficking and multiple handling. Requirements for surplus topsoil *are to be set out in Appendix 6/1. The Sixth Edition required it to be stockpiled in locations and in the manner described in the Contract — NG 602 states that it is to be stored wherever practicable*. Storage areas should be shown on the drawings and detailed in Appendix 6/8.

Excavations for foundations and trenches (*pits omitted*) are to be adequately supported and may not be battered unless otherwise detailed in Appendix 6/3. If they are battered they should be benched as described in Appendix 6/3 prior to backfilling and compaction. The additional work and material are to be provided by the Contractor (presumably item coverage 18(g) and 18(k) are intended to cover this work).

The Contractor must keep earthworks free of water (but see Box 11), including rapid removal of water, lowering water level in excavations, forming and shaping earthworks, pumping and temporary watercourses.

Material within 450 mm (*or 350 mm if the annual frost index of the site is less than 50*) of final road surface *or paved central reserve* subject to tolerances shall be non-frost susceptible. *The details of the testing have been changed. Neither the Notes for Guidance nor the Sample Appendix refer to the annual frost index which could lead to ambiguities. It would be preferable to clearly state what the designer's requirements are in this respect.*

603 • Forming of Cuttings and Cutting Slopes

Cuttings shall be excavated to the lines and levels *described in Appendix 6/3*. Undercutting of slopes and toes of cuttings is to be restricted to that required by the Contract and any restriction on extent should be shown in Appendix 6/3 (*see Boxes 7 and 12 for measurement aspects of excavation.*)

Itemisation

16 Separate items shall be provided for excavation in accordance with Chapter II paragraphs 3 and 4, and the following:

Group	Feature	
I	1	Excavation.
II	1	Acceptable material Class 5A.
	2	Acceptable material Class 3.
	3	Acceptable material excluding Classes 3 and 5A.
	4	Unacceptable material Class U1.
	5	Unacceptable material Class U2.
III	1	Cutting and other excavation.
	2	Structural foundations.
	3	Foundations for corrugated steel buried structures and the like.
	4	New watercourses.
	5	Enlarged watercourses.
	6	Intercepting ditches.
	7	Clearing abandoned watercourses.
	8	Removal of surcharge.
	9	*Gabion walling and mattresses.*
	10	*Crib walling.*
	11	*Caps to mine working, well, swallow hole and the like.*
IV	1	0 metres to 3 metres in depth.
	2	0 metres to 6 metres in depth and so on in steps of 3 metres.

Note 1: Acceptable material Class 5A shall not be separately identified by any Group III or IV Feature.

Note 2: Group IV Features shall be applied only to Features 2, 3, *9, 10 and 11* of Group III.

Box 12 MMHW — Excavation Itemisation — III Features 9, 10 and 11 introduced through October 1991 Addendum

An example of a completed Appendix 6/3 is included in Annex A

Box 13

Excavations can be halted at any stage provided 300 mm is left as weather protection to the formation or sub-formation. If requirements are otherwise they must be specified in Appendix 6/3.

Requirements for pre-split blasting should be shown in Appendix 6/3 and details are to be supplied to the Engineer for approval (see also Box 18).

The specified treatment of faces which are not to receive topsoil includes cleaning by airline, removing soft and insecure material, replacing with concrete *mix ST2 (Sixth Edition C10P)*, trimming back to make stable, sprayed concrete, reinforcement, weepholes, netting, rock bolts, etc. These are either detailed in SHW, set out in Appendices 6/3 or 6/10, or required by the Engineer. *The Sixth Edition Notes for Guidance stated that these details were to be shown on drawings. They are now to be described in Appendix 6/3.* If the details are required by the Engineer without being stipulated in Appendix 6/3 or if the details change then, clearly, the rates cannot have allowed for this work and there are no appropriate measurement items in the MMHW except item coverage (see Box 14) which seems inappropriate for such detailed work.

Excavation of Acceptable Material Class 3, and Acceptable Material Excluding Classes 3 and 5A

Item coverage

(m) treatment of faces of cuttings which are not to receive topsoil;

Box 14 MMHW para. 18

Faces which are to receive topsoil must have soft areas removed and either replaced with similar fill or concrete which should match surrounding material. Requirements are to be shown in Appendix 6/3 or as required by the Engineer.

604 • Excavation for Foundations

The bottom of all foundation excavations shall be formed to lines and levels *shown on drawings*. Blinding concrete should also be *shown on drawings. The Sixth Edition SHW wording was 'described in Contract', although the Notes for Guidance indicated it should be shown on drawings*. Drawings should show where battering of slopes is allowable *and details given in Appendix 6/3.*

Soft spots are to be backfilled with concrete *mix ST1* or other material required by the Engineer (class 6K material is to be used for corrugated buried structures). Measurement will be as a soft spot only if it is less than 1 m^3.

605 • Special Requirements for Class 3 Material

Para 18 Item Coverage

(u) complying with special requirements for Class 3 material and other materials requiring special treatments.

Box 15 MMHW — Excavation Item Coverage

Chalk which could be degraded by normal earthmoving operations should be designated as Class 3 material even though it may include harder chalks or other material (see Box 16). Class 3 material and the restrictions imposed by this clause only apply if the material is designated as such in Appendix 6/1 and referred to the drawings for extent. There is no provision for the Engineer to re-classify material as Class 3 during construction. Notes for Guidance 605.2 states 'otherwise chalk for general fill is dealt with as a Class 1 or Class 2 material as appropriate'. Thus imported material consisting mainly of classifications C and D would not have been designated as Class 3 and there seems no provision for dealing satisfactorily with this situation.

5.14 The time period, when no Class 3 earthworks shall take place (Clause 605.1(i)) ie Winter. This is intended to be the period when the rainfall exceeds the evaporation rate, over monthly or weekly periods. . . . A practical approach to the prediction of the most likely period is to use the rainfall and the evaporation figures from the Meteorological Office long-term statistical records. Analyses of records for the last 25 years are available for any area in the UK and may be obtained from the Meteorological Office Advisory Service at Bracknell. If the appropriate rainfall or evaporation figures are not available, then a winter period of 1 November to 31 March should be specified.

5.15 Minimum height of excavation Clause 605.1 (iii). If possible this height should be extended to 5 m and worked as a quarry face. However it would be preferable to work a 6 m high cutting in two layers of 3 m each rather than a 5 m layer followed be a 1 m layer, and so the depth of chalk in the cuttings must be considered before amending the minimum excavation height. If the chalk is surplus and will be taken off-site, then the plant restriction and minimum height may be ignored.

5.16 If the majority of the chalk is estimated to be Class A or B the restriction of haulage vehicles to a 15 m^3 maximum capacity (Clause 605.1 (v)) may be relaxed. If there is any doubt it would be preferable to leave that decision to the Engineer on site after close observation of the behaviour of the material.

5.17 Layering of Class 3 Chalk with other materials is not recommended unless absolutely unavoidable (Clause 605.1 (vii)). However if chalk is placed on a less permeable material, that material must be cambered to shed water transversely. Failure to do so could cause free water released during the placing and compacting processes to pond inside the chalk with possible long term instability setting in. Therefore if composite embankments are envisaged the surface of any relatively impermeable material must be cambered, and a suitable requirement to this effect included in Appendix 6/4. For the same reason Class 3 material must extend the full width of the embankment and must not be contained within bunds of impermeable material.

5.18 Period of waiting (Clause 605.1) If required, a pre-contract trial may be carried out on the chalk to investigate it's readiness to become unstable and then stable again. If the time for the hardening process is delayed beyond 4 weeks the period in Appendix 6/4 should be lengthened accordingly: similarly if the chalk hardens quickly the period may be reduced.

Box 16 Information for Appendix 6/4 — HA 44/91

Box 17 MMHW — Excavation of Watercourse -- 90 and 92 changes introduced in Addendum

> 15 The measurement of excavation shall be, for:
>
> (f) New and enlarged watercourses, intercepting ditches — the volume of the void formed from Existing Ground Level down to the outline stated in the Contract less the volume of topsoil Class 5A in the void included in the measurement under paragraph 15(a) of this Series.
>
> (g) Clearing abandoned watercourses — the volume of the void formed from Existing Ground Level down to the outline stated in the Contract.
>
> 90 *Measurement of bagwork shall be the flat undeveloped area of the work.*
>
> 92 *(g) bags, filling, staking and securing.*

Appendix 6/4 should state the period during which excavation cannot take place (see Box 16) although the Engineer may exceptionally agree otherwise (such a relaxation would have payment implications). There is a restriction on excavation plant to minimise degradation and on layering with other materials. These cannot be stockpiled or multiple handled.

Material other than chalk, within the designated Class 3 material, shall be excavated and deposited separately from chalk. Although it is not chalk, this material has been designated as Class 3 and will be measured as such.

Appendix 6/4 should specify details of minimum face height, trafficking, filling requirements, sealing of material and *any other special requirements* (see Box 16).

The Contractor must delay depositon if instability is occurring due to excessive working or inclement weather.

606 • Watercourses

Requirements are to be set out in Appendix 6/3. Any special treatment to the surface of filled redundant watercourses *should be described in Appendix 6/3 (see Box 17 for changes in measurement).*

Excavated material is to be dealt with as unacceptable material.

New and cleared watercourses are to be maintained in a clear condition. There is no stipulation as to what period of time applies.

607 • Explosives and Blasting for Excavation

Blasting is only permitted if it is noted in Appendix 6/3 with limits on locations and times (see Box 18) and no plaster shooting is allowed. NG 607.1 requires pre-construction consultation before allowing blasting in Appendix 6/3.

The Contractor has to obtain the written consent of the Engineer at least 10 days before the programme commences and give written notice 12 hours before each event. There is no stipulation as to how long the Engineer has to consider and give consent.

> 7.14 The term 'explosives' includes both high explosives and also slow-burning materials such as propellants. Hence devices incorporating a cartridge resembling a shot-gun cartridge for boulder or rock splitting are also covered by SHW Clause 607.
>
> 7.15 The use of blasting for excavation should only be allowed in SHW Appendix 6/3 where it is considered to be necessary. Where it is considered not likely to cause damage or nuisance, the contractor should be permitted in Appendix 6/3 to use it as an alternative method of excavation.
>
> 7.16 To avoid intrusion, permitted hours for blasting entered in Appendix 6/3 should not extend outside normal working hours except where the site is remote from any inhabited area.

Box 18 HA 44/91 — Explosives

The Contractor has to fix instruments, making his own arrangements for property off site, and take measurements and report results to the Engineer each day. Noise must comply with SHW Clause 109.

There are minor layout changes to this clause and 'subsequent amending Regulations' added to (2)(xii) and BS6657 in (2)(xiv).

608 • Construction of Fills

Material is to be deposited in layers to meet the requirements of Clause 612. Material to be end-product compacted must be deposited in layers not exceeding 250 mm (see Box 19 for MMHW rules on 'Deposition'). Only Class 6A material can be deposited into open water, by end-tipping without compaction (see Box 24 for measurement rules).

Drawings should show the locations and requirements for selected fills, staged construction, surcharging, benching and each change of cross-section of fill (Notes for Guidance 608.1–608.4).

Plant movement over starter layers may need to be restricted. If necessary to prevent damage, the Engineer can reduce passes or require lighter plant.

Class 1C and 6B material (coarse granular) shall be spread by crawler tractor and blinded if voids remain. *'Compaction' item coverage has now been amended to include this requirement (see Box 30).*

SHW 608.4 states that embankments shall be constructed to their full

30 The measurement of deposition of fill shall be the volume of compacted fill, calculated in accordance with paragraphs 47, 48 and 49 of this Series, less the volume of imported fill calculated in accordance with paragraphs 41 and 42 of this Series.

32 Separate items shall be provided for deposition of fill in accordance with Chapter II paragraphs 3 and 4 and the following:

Group	Feature	
I	1	Deposition.
II	1	Acceptable material.
	2	Acceptable material Class 1C.
	3	Acceptable material Class 3.
	4	Acceptable material Class 6B.
III	1	Embankments and other areas of fill.
	2	Strengthened embankments.
	3	Reinforced earth structures.
	4	Anchored earth structures.
	5	Landscape areas.
	6	*Environmental bunds.*
	7	Fill to structures.
	8	Fill above structural concrete foundations.
	9	Fill on sub-base material, roadbase and capping.
	10	Fill on bridges (under footways, verges and central reserves).
	11	Upper bedding to corrugated steel buried structures and the like.
	12	Lower bedding to corrugated steel buried structures and the like.
	13	Surround to corrugated steel buried structures and the like.
	14	*Fill above corrugated steel buried structures and the like*

Note that deposition is not measured on imported material although HA 44/91 para. 3.46 implies otherwise

Box 19 MMHW — Measurement and Itemisation of Deposition

width unless otherwise required. Notes for Guidance state that drawings should show details *cross-referenced to Appendix 6/3*. This will often be necessary if the embankment is being constructed alongside existing traffic and similarly it should be shown if steeper side slopes permitted.

SHW 608.9 still requires that the last 600 mm of fill must be completed in a continuous operation over the full width which will be impractical in the situation instanced above and conflicts with the requirements of SHW 608.4. *This ambiguity could very easily have been resolved in the Seventh Edition.*

If the Contractor does not form the formation or sub-formation after completion of the last 600 mm he must without delay place a minimum of 300 mm of the same material as the formation as weather protection. It must be constructed in a continuous operation and provided from the Contractor's own resources. The Engineer's approval is required for surcharge material or it should be described in Appendix 6/3. If it settles below the formation or sub-formation it may need replacing with acceptable material. Surcharge material is separately measured under 'excavation' but not under 'deposition' and 'compaction' items (see Boxes 12 and 19).

PFA is to be protected from erosion during construction and completed slopes covered immediately with topsoil or turf as required in the Contract.

Side slopes of embankments in cohesive material are to be sealed by tracking (this point is covered in the item coverage for compaction).

609 • Geotextiles Used to Separate Earthworks Materials

As with the general philosophy of the Seventh Edition, clauses now refer to Appendix 6/5 rather than as described in the Contract. The designer, through Appendix 6/5, should state the location, sampling, lapping and life required (see Box 20).

The Contractor must provide evidence that the material will be sufficiently durable so that its integrity will be maintained for life even though NG 609.1 states that there are no tests acceptable for durability. If specifying geotextiles for purposes other than separation of earthworks materials, the tests may need modification. Samples should be jointly selected (the number being described in Appendix 6/5) and tested to show they meet the requirements of Sub-clauses (4)(i) and (ii). They must also be retained by the Contractor until the end of the Maintenance Period. *The testing details have been changed by omission of most of Sub-clauses 8, 9 and 10 and by reference to BS 6906. Surprisingly, there is no reference to Appendices 1/5 or 1/6 which should list all tests or samples the Contractor is expected to provide.*

Separation

5.81 Geotextiles may be used to separate two different materials which would otherwise have a tendency to mix under working conditions.

There is also an added function whereby the geotextile being permeable, allows the passage of water across it to reduce the build-up of undesirable pore water pressures in either of the adjacent materials. Clause 609 SHW defines some of the minimum physical properties that the geotextile must possess. If different values of these, or other, properties are required then they should be inserted in Appendix 6/5. The NG Sample Appendix 6/5 gives details of the information to be included. The amount of lapping of the geotextile is specified as 300 mm minimum, but the designer should consider the maximum possible amount of differential settlement and deformation that could occur in the soil close to the geotextile. If the differential settlement is more than 200 mm the minimum lap stated in Appendix 6/5 should be increased accordingly. In normal circumstances, the geotextile should not be physically jointed but allow the material to slide and not 'bridge' over low areas where it could become overstressed and tear.

Box 20 Geotexiles — HA 44/91 Chapter 5

There should be no sharp projections of the underlying layer, the geotextile must be in continuous contact, not stretched over hollows and covered immediately.

610 • Fill to Structures

This clause excludes fill to reinforced earth, anchored earth and corrugated steel buried structures which are covered in separate clauses.

The selection of appropriate fill material will depend on design assumptions and any restrictions should be stated (Classes 6N, 6P, 7A or 7B used in locations described in Appendix 6/6). With 7A there is a risk of swelling so the maximum and minimum values of m/c, MCV and shear strength should be stipulated as well as the plasticity limits of Table 6/1 (see also Box 2).

End product compaction is specified but with restrictions within 2 m of the structure and maintaining a level within 250 mm of fill outside the zone. Fill must be brought up either side evenly. *This applies to a structural element or a buried structure except where Clause 623 applies.* A test for stability may be required in Appendix 6/6 *but this will also need to be listed in Appendix 1/5.*

> An example of a completed Appendix 6/6 is included in Annex A

Box 21

611 • Fill above Structural Concrete Foundations

Full details are to be shown on the drawings (see Box 23). Previously these were to be shown in Appendix 6/6.

The fill is to be specified in Appendix 6/6 as either 6N, 6P, 7A or 7B or other selected or general fill complying with Table 6/1 — subject then to the restrictions of SHW Clauses 610.4 and 610.5 (see Box 22).

612 • Compaction of Fills

> Para 5.30 HA 44/91
>
> In the unburnt state, minestone is not one of the materials for the selected fills defined in Classes 6 & 7 of Table 6/1 SHW: the use of minestone as a fill to bridge abutments and retaining walls is also prohibited in BD 30/87.

Box 22 Minestone restrictions

Compaction should be carried out as soon as practicable after deposition. Table 6/1 specifies the method and Table 6/4 gives the details (see Boxes 24 and 25 for measurement rules). Dynamic compaction is separately covered in Clause 630.

Method compaction will be used for the majority of fills and Table 6/4 should achieve a minimum degree of compaction equivalent to 10% air voids (see Box 26).

The Contractor must apply in writing for the Engineer's permission 24 hours before he intends to carry out compaction outside normal hours but there is nothing to cover a refusal by the Engineer.

If the Contractor wishes to use plant and methods not in Table 6/4 he must demonstrate at the site trial that they are equivalent. *The item coverage used to include 'compaction trials and demonstrations', but it has now been amended to 'trial areas, trials and demonstrations'.*

The Engineer may carry out field dry-density tests and compare these with approved work. If compaction is inadequate due to failure of the Contractor, he must carry out further work *as is required to comply with the Contract (the reference to 'as the Engineer may decide' is omitted).* If compaction is found to be inadequate but the Contractor has complied with the Specification, then the Engineer will need to consider an ordered variation.

The sub-clause setting out the rules to apply to Table 6/4 has been re-arranged but with no change to the requirements. For the purposes of Table 6/4:

The number of passes is doubled on some materials within 600 mm of formation.

The requirements for tamping rollers apply to machines having two rolls in tandem; if there is only one roll, then the number of passes should be doubled.

When vibrating rollers are used, the lowest gear should be selected

on self propelled rollers or a speed limit of 1.5–2.5 km/h for a towed machine should be imposed; if higher speeds are used the number of passes should be increased.
A tandem vibrating roller requires half the number of passes for a single roll.
The frequency and speed of vibrating rollers must be capable of being read by an inspector alongside the machine.
With a combination of plant types the depth of layer shall be the minimum for the plant and the number of passes shall be the greatest.

End product compaction is restricted to PFA in general fill and the Contractor must supply information at least seven working days before commencement. The fill must attain the percentage of maximum dry density approved by Engineer stated in Table 6/1 and density testing will be used to check compliance. *Nuclear methods of measuring field dry density may be used if required in Appendix 6/3 or permitted by the Engineer. This sub-clause is considerably changed and the use of nuclear gauges must comply with Clause 123.*

Box 23 HA 44/91 Chapter 3

3.41 The earthworks to structures drawing (E) should indicate the extent of filling to each structure and fill above structural concrete foundations in order that the quantity and location of each class of material may be determined. The actual quantities should be included on the earthworks drawings' longitudinal sections (C) along with all the other quantities.

Box 24 MMHW Measurement of Compaction

47 The measurement of compaction of fill in embankments and other areas of fill, in strengthened embankments, in reinforced earth structures, in anchored earth structures, in landscape areas and in *environmental* bunds shall be the volume of the embankment or void filled from Existing Ground Level up to the Earthworks Outline plus, where required by the Contract, the volume of:

(a) the void formed by the removal of topsoil Class 5A beneath the fill in question, and included in the measurement under paragraph 15(a) of this Series;
(b) the void formed by excavation for the fill in question
    (i) below the Earthworks Outline included in the measurement under paragraph 15(b)(i) of this Series; and
    (ii) below Existing Ground Level included in the measurement under paragraph 15(b) (ii) of this Series;
(c) surcharge, being the void filled from the Earthworks Outline up to the profile stated in the Contract to which the surcharge is required to be constructed;

less in each case the volume of any compaction of fill to structures, and bedding and surround to corrugated steel buried structures and the like included in the volume so obtained and which is measured separately under paragraph 49 of this Series.

Note that the measurement of compaction will apply even to those materials not being compacted although HA 44/91 para. 3.46 implies otherwise

Box 25 Clarification of item coverage

A major source of difficulty in the Third Edition MMHW was the short-hand referencing of the item coverage. Thus the import item coverage read:

(a) compaction of fill (as this section paragraph 45(a) to (g) inclusive)

Understandably many compilers not familiar with the detailed measurement rules assumed this meant compaction was not measured on import which resulted in an omission from the Bill of Quantities (see also Box 24).
This has now been overcome by simply listing the item coverage in full.

4.40 Two basic types of compaction specification can be applied.

4.41 Method. The method of compaction is specified in terms of plant, method of operation, thickness of layer and number of passes. This type of compaction specification is detailed in Table 6/4 of the SHW and is the type specified for most of the material classes in Table 6/1 of the SHW. The compactive effort specified in Table 6/4 for Methods 1 and 2 is designed to produce a maximum air-voids content of 10% assuming a conservative moisture content which is dry of the average field moisture content for the relevant material.

4.42 Method 3 should give approx 95% BS 1377: Part 4 (2.5 kg rammer method), maximum dry density assuming a conservative moisture content; Method 4 should produce a maximum 10% air voids at high moisture content (chalk); Method 5 is based on continental experience which gives satisfactory performance when this compactive effort is used on coarse granular material; Method 6 should produce a maximum 5% air voids at a lower limit of moisture content for sub-base compaction, and Method 7 should produce a maximum 5% air voids at a MCV of 12. However this does not preclude provision being made for some insitu density testing to ensure the compactive effort is sufficient. Where a material lies across a boundary between two of the different methods of compaction described in Table 6/4, then the method which requires the higher compactive effort should be specified.

4.43 End Product. The state of compaction to be achieved is specified, leaving the choice of method by which it is achieved to the Contractor. Even so, it may be advisable in some instances in Appendix 6/3 of SHW to restrict the thickness of each compacted layer so that effective control can be maintained on site. This control will necessitate the determination of the insitu bulk density and moisture content for the complete depth of the compacted layer or, failing this, for the lower 150 mm of the compacted layer, together with any other material properly defined in Table 6/1 of SHW.

Box 26 Compaction — Method and End Product (HA 44/91)

613 • Sub-formation and Capping

The Specification does not require a minimum CBR of formation or sub-formation (except in the case of stabilised capping) but permitted capping materials should result in a CBR of 15% (see Box 27).

The drawings should show the location and thickness of capping including areas where the sub-formation may have a different fall to the formation above it. Capping may be of one class or two elements of different classes and permitted options for classes of capping should be described taking account of the properties of the likely sub-formation (see Box 27).

Appendix 6/7 should also show minimum thickness for protection where the fill characteristics do not require the full thickness immediately. Appendix 6/7 should show any requirement for a demonstration area and the Contractor shall demonstrate to the Engineer the methods, materials and equipment he proposes to use by constructing such an area. Suitable locations should be made available so that it can form part of the permanent works.

10.24 Capping materials described in SHW are designed to provide a stiffness and strength equivalent to a CBR of at least 15% at their top surface when placed and compacted to the thickness commensurate with the sub-grade CBR given in Table 10/1. SHW Table 6/1 Class 6F1 and 6F2 are selected granular materials of fine and coarse grading respectively . . .
The Designer should allow the widest possible choice of capping materials to be used and should only restrict the choice if there are sound engineering reasons.

Box 27 Capping Materials — HA 44/91

Box 28 Compaction — Changed item coverage — (e), (m) and (o) introduced in Addendum

- (e) *complying with the requirements for Class 3 material and other materials requiring special treatment;*
- (i) *trial areas, trials and demonstrations;*
- (j) *awaiting Engineer's approval of trial areas, trials and demonstrations;*
- (k) *making good after sampling and testing;*
- (m) *treatment of side slopes and berms;*
- (o) *blinding.*

The Contractor must allow not more than five working days after acceptance by the Engineer to enable the Engineer to carry out tests on each demonstration area (what these tests are is not stated and it is not readily apparent, as method compaction is used and no CBR's are specified). This sub-clause (5) is badly drafted and has led to widely different interpretations with the Sixth Edition *but is unchanged in the Seventh Edition.*

The Fourth Edition MMHW attempts to overcome at least some of the problems by introducing additional item coverage into the compaction item (see Box 28) although it would have seemed more approppriate in the imported fill item coverage. The MMHW now requires the Contractor to allow for 'trial areas, trials and demonstrations;' 'awaiting Engineer's approval of trial areas, trials and demonstrations;' and 'making good after sampling and testing' but the central question of whether the Contractor has to await the results of the Engineer's tests is still unclear.

The Contractor shall limit the areas of unprotected sub-formation to suit the rate of deposition and unprotected sub-formation must be covered overnight (unless permitted otherwise by the Engineer) and protected from rain.

Appendix 6/7 should permit or require treatment of sub-formation in cuttings in line with the alternatives in Sub-clause 11, similarly for embankments in accordance with the alternatives in Sub-clause 12.

If a stabilised layer is overlain by 6F1 or 6F2, the stabilised layer shall be compacted as sub-formation. The Sixth Edition had this requirement in respect of embankments but not for cuttings.

614 • Cement Stabilisation to Form Capping

Cement stabilisation of Class 6E granular, 7F silty cohesive or 7G PFA should be included as an option if pre-Contract trials show it to be feasible. There is no indication of how such an option should be allowed for in the documents — presumably by allowing both conventional and cement stabilisation but the measurement rules do not cover this (see Box 32).

Box 29 MMHW — Completion of formation and sub-formation

Group II Features — 2 and 3 are new

1 On acceptable material
2 On Class 1C material
3 On Class 6B material
4 On rock in cuttings

Item coverage

- (a) removal of protective layer, mud and slurry;
- (b) compaction;
- (c) cleaning, trimming, regulating, making good and rolling;
- (d) cement bound materials;
- (e) excavation, processing, compaction of naturally occurring Hard Material;
- (f) measures to protect formation and sub-formation against deterioration or degradation.

10.68 The grading envelope for Class 6E material comprises essentially granular materials for cement stabilisation whereas Class 7F material has been introduced so as to include the silty sands, sandy silts and mixtures of sand, silt and clay, all capable of being stabilised with cement.

10.69 The minimum bearing ratio to be inserted in Appendix 6/1 for classes 9A, 9B and 9C will normally be 15% at 7 days. This is the strength required in the short term during construction when the capping has to act as a 'platform'. In the long term a design bearing ratio of 7–8% will allow for possible softening effects. A cement content of 2% added to Class 6E material will normally give the required 7 day strength for Class 9A mateial, but this should be checked on the demonstration area. Class 7F and 7G materials will in general require larger additions of cement to produce Class 9B and 9C cappings respectively. In nearly every case there should be a requirement in Appendix 6/7 for the construction and testing of a demonstration area, unless pre-Contract trials have already been carried out. Except for uniformly graded materials, the strength of a cement-stabilised material is directly proportional to cement content.

10.74 It is important to maintain close site supervision to ensure that the depth processed and the final compacted thickness are correct, particularly for layers thicker than 250 mm.

10.80 Although soil-cement CBM 1 for sub-bases and roadbases (SHW Clause 1035) is generally non-frost susceptible, the same is not always true for capping where the cement content is lower. In particular, stabilised PFA (Class 9C) is known to be frost susceptible. All cement stabilised materials should be tested after 28 days curing using the procedures in Clause 602.19 SHW. Weather protection during the winter months, as described in Paragraphs 10.53 to 10.57, should be provided for doubtful Class 9B materials and all Class 9C materials.

Box 30 Cement Stabilisation — HA 44/91

The amount of cement to be added to the material will be 'that quantity of cement measured as a percentage of its dry weight as determined on the demonstration area to meet the required bearing ratio in Appendix 6/1'. If there are doubts about the sufficiency of the 2% minimum of cement, the results of trials should be made available to the tenderers, reference being made to them in Appendix 6/7 (see Box 30).

It is not at all clear what is being 'determined' on the demonstration area and by whom. *The Advice Note refers to the cement content being 'checked' on the demonstration area (see Box 30) which would imply the possibility of a change in cement content but the MMHW does not allow for this and the basis of the Contractor's price will be uncertain as it was with the Sixth Edition.* Unlike the demonstration area for capping in SHW Clause 613.5, there is no statement that, if successful, it can remain as part of the Permanent Works and neither is there a requirement for its removal as in SHW Clause 613.

The 'compaction' item coverage has been amended, presumably to cover the cost of this demonstration area (see Box 28) although it would seem more appropriate to the Soil Stabilisation item.

The October 1991 Addendum introduced a new bill item for 'Proof Demonstration of Capping Construction' in accordance with the requirements of Clause 36(3) of the ICE Conditions of Contract but this does not appear in the Fourth Edition. Instead, the MCD amends the ICE Clause (see Box 31).

The rate of spread of cement must be checked by the Contractor once every 500 m^2 in the presence of the Engineer (there is no mention of the Engineer's Representative as an alternative).

Stabilising is by pulverising and mixing with cement until 95% of the clay and silt fraction passes a 28 mm sieve. The material is to be stabilised in a single layer if its compacted thickness is 250 mm or less. If the

Box 31 MCD changes to ICE Clause 36(3)

> **Clause 36 (3) is deleted** — the following is part of the new sub-clause and the reference in the original clause to proof of design tests disappears
>
> **Tests**
>
> (3) Each test of materials or workmanship which is specified in the Contract as to be carried out by the Contractor shall be carried out by him at his own cost provided that the test in question is particularised in the Specification in sufficient detail to enable the Contractor to have priced or allowed for the same in his Tender . . .

compacted thickness is greater (which requires the Engineer's approval) then it is stabilised in layers not less than 130 mm or greater than 250 mm thick. An overlap of 150 mm between adjacent passes is specified and it must cut into the lower layer by at least 20 mm. If processing a layer on top of a lower, previously stabilised layer, the blades must cut into it by at least 20 mm.

Method compaction is used for Classes 9A and 9B (in accordance with Table 6/4, unless layers are greater than 250 mm when the number of passes depends on the demonstration area results) and end product compaction is used for Class 9C.

Class 9B must have a MCV not greater than 12 nor less than the figure in Appendix 6/1 before compacting and compaction must be complete within two hours of mixing (Sample Appendix 6/1 includes a minimum MCV for lime stabilised material but not cement.)

Curing must be in accordance with Clause 1035. The materials shall be protected to prevent freezing for seven days during periods when low temperatures or ground frost are forecast. Further work above the material is not permitted until bearing ratio stated in Appendix 6/1 is reached.

615 • Lime Stabilisation to Form Capping

Unlike cement stabilisation there is no advice in the Notes for Guidance to include this treatment as an option. The measurement rules are the same as with cement stabilisation (see Box 32).

Lower value of MCV for Class 9D and of mc or MCV for Class 7E is to be specified in Appendix 6/1 after determining it from laboratory tests (NG 615.1). Sub-clause 13 requires the upper MCV to be stated for 9D material. Appendix 6/1 should also include the required bearing ratio.

Appendix 6/7 should specify either quicklime or hydrated lime and any additional tests for rate of spread. Notes for Guidance Clause 615.1 states

Box 32 Soil Stabilisation — MMHW

> 54 The measurement of soil stabilisation shall be the volume of the material to be stabilised measured to the outlines stated in the Contract irrespective of the number of layers or thicknesses, methods or sequences of operations involved in stabilising the material to the depth required.
>
> Note: Soil stabilisation means the process of stabilisation whether the material is intact and undisturbed or deposited and compacted prior to stabilisation.
>
> Excavation, fill, import, disposal, deposition and compaction required to expose or produce the layer to be stabilised, as appropriate, shall be included under the measurement of earthworks elsewhere in this Series.
>
> Excavation, deposition and compaction involved in the process of stabilisation itself shall not be measured.
>
> Group III Features 1 Cement.
> 2 Lime.

that $2\frac{1}{2}$% of available lime will give sufficient long-term strength for a capping but the actual weight of lime as a percentage is determined from the demonstration area.

As in SHW Clause 614, the wording regarding the significance of the demonstration area is not clear. Similarly, the specification does not state whether the demonstration area can be incorporated into the works (as specifically allowed in SHW Clause 613.5) and there is no requirement for its removal as in Clause 613.

The Contractor checks the rate of spread in the presence of the Engineer (no mention of the Engineer's Representative) and tests for available lime content. Sub-clause 6 refers to 'or other rate of spread as described in Appendix 6/7' which makes no sense. It was presumably meant to refer to 'other rate of testing' which may be specified in Appendix 6/7. *This error remains uncorrected in the Seventh Edition.*

Stabilisation is limited to March–September and a shade temperature of 7°C or above unless a lower figure has been agreed from the demonstration area.

Layers must not be less than 130 mm or more than 250 mm thick, must overlap by 150 mm and cut into the underlying layer by a minimum of 20 mm. There must be at least two passes of the machine and 95% of the resultant material must pass a BS 28 mm sieve (see Box 33).

The surface is then sealed by rolling, after which the processing must be interrupted by at least 24 hours and not more than 72 hours (to be agreed by the Engineer) for the lime to react. As least one further pass is required to ensure compliance with Sub-clauses 9 and 13 followed by compaction.

Method compaction is to Table 6/4 Method 7, but processed material must have a MCV within the range specified in Appendix 6/1 (*unlike Clause*

Box 33 Lime Stabilisation — HA 44/91 Chapter 10

10.36 A bearing ratio of up to 8% can be expected immediately following final compaction, equivalent to an upper limit for MCV of 12. The addition of a minimum of $2\frac{1}{2}$% available lime should result in the CBR rising to a minimum value of 15% in the short-term after compaction and it should not fall below a minimum value of 7–8% in the long-term. This allows for some degree of softening which may occur.

10.39 In order to perform to the SHW, the 'approved spreading machine' referred to in Clause 615.6 is likely to be one with a device capable of accurately metering the lime supplied, gearing to allow for variation in speed of travel, large capacity and spreading width, and provide precautions against dusting.

10.40 Clause 615.10 SHW requires the stabilising machine to have an integral spray bar. This is to ensure that water can be added evenly at the first or second pass of the machine which is particularly important when quicklime is used. Machines which do not incorporate their own integral water supply are unlikely to be able to properly operate the integral spray bar specified in SHW.

10.41 It is important to maintain close site supervision to ensure that the depth processed and the final compacted thickness are correct, particularly for layers thicker than 250 mm.

10.47 Where the Engineer agrees that layers more than 250 mm thick can be constructed, the Contractor is required to demonstrate that such an operation is feasible with the plant he proposes to use.

10.55 Lime stabilised cappings should be covered during the period October to February by a weather protection at least 300 mm thick if the overlaying pavement has not been constructed, unless the result of the BS 812 tests demonstrate that the stabilised material is non frost-susceptible. This should be established at the ground investigation stage so that if weather protection is required it can be included in the Contract. The sub-base is not sufficient as weather protection if it is less than 300 mm thick.

614.11 there is no provision for layers in excess of 250 mm although HA 44/91 para. 10.47 implies that the Engineer can agree to this — see Box 33). If compaction is delayed then the surface is sealed and reprocessed before commencing compaction.

The required bearing ratio must be obtained before further work is carried out on the material.

616 • Preparation and Surface Treatment of Formation

Preparation of formation requires removal of any protective layer, removal and replacement of any soft or damaged areas. The compaction shall be that for a 250 mm thick layer (in addition to any compaction required for fill) and trimming shall be carried out to tolerances within +20 mm and −30 mm.

If tolerances are exceeded the Contractor must re-trim if too high and add material if too low. If the trimming is too low by more than 150 mm with Class 2 and 7 materials, he must excavate and refill.

After trimming it is to be rolled with a single pass of a smooth wheeled roller or (except for Class 3) a vibrating roller or vibrating plate compactor. Sub-grade drainage should be complete before laying the sub-base on the completed formation.

The Contractor must limit the preparation of the formation to suit his rate of deposition of sub-base and Classes 2 and 7 materials must be covered overnight unless permitted otherwise by the Engineer.

Where required in Appendix 6/7 formation on rock should be processed as set out in the Appendix or, if the rock is tabular, regulated using cement bound material or concrete *Mix ST1*. In either case the Engineer can require and agree the treatment if it is not shown in Appendix 6/7.

The Contractor is meant to include for these costs in his rates as the item coverage for 'completion of formation and sub-formation' includes (c) cleaning, trimming, regulating, making good and rolling; and (d) cement bound materials. This must be interpreted having regard to the basic principles of the MMHW including the assumption that full construction requirements are given.

617 • Use of Sub-Formation or Formation by Construction Plant

There is a minor re-arrangement of phrases and the removal from SHW to NG of the Engineer permitting sub-base traffic to use formation or capping.

There is a restriction on trafficking sub-formation and formation unless they are adequately protected, if necessary in addition to weather protection. The Contractor has to submit proposals for approval in areas where sub-formation or formation are within 300 mm of ground level.

618 • Topsoiling, *Grass Seeding and Turfing*

Appendix 6/8 should restrict excavation from topsoil stockpiles in line with Sub-clause 3 if the soil is likely to have a high clay content (see Box 35).

Hydraulic mulching is allowed as Treatment III and it may be used for seed and fertiliser on topsoil in Treatment 1 unless otherwise stated in Appendix 6/8 (see also Box 35). Mulch has to be approved by the Engineer and a retaining agent may be stipulated.

Cutting slopes must be benched where required in the Contract and harrowed diagonally before topsoiling. Any areas of cutting slopes which do not need harrowing should be listed in Appendix 6/8, also if the depth is not 50 mm. *There was an omission in the item coverage in the Third Edition of the MMHW regarding this requirement. This has now been corrected.*

Topsoil should not be spread by tracked vehicle if specified in Appendix 6/8. *The Sample Appendix now lists this as an option. It was omitted from*

> An example of a completed Appendix 6/8 is included in Annex A

Box 34

> 12.5 It is usual for the earthworks drawings to show the extent and depths of topsoiling. The drawings will also show the treatment type in accordance with SHW Clause 618.4. Other details on the treatment of topsoil should be set out in SHW Appendix 6/8.
>
> 12.7 The requirements of SHW Clause 618.3 only apply when the sub-soil is a heavy clay. Where this occurs then the extent of the restriction will need to be set out in SHW Appendix 6/8. The rainfall figure quoted in SHW Clause 618.3 is satisfactory for most soils likely to be encountered in this country.

Box 35 HA 44/91 — Topsoil

the Sixth Edition Sample Appendix. Strangely the item coverage includes compaction.

Topsoil should have its upper 50 mm reduced to a fine tilth, have fertiliser raked in, be seeded and immediately lightly raked. Appendix 6/8 should stipulate topsoil thickness and the rate of spread of seeding and fertiliser if different from the specification. *The rates of spread of seeding have been changed from the Sixth Edition requirement of not less than 30 g/m² for side slopes and 20 g/m² elsewhere to 20 g/m² and 10 g/m² respectively.*

Turf may be cut from within the site (with the permission of the Engineer) from those areas of Class 5A material required to be stripped and must be retained as described in Appendix 6/8 and be regularly watered.

Areas which are not required to be mown or are to be mown three times should be detailed in Appendix 6/8, otherwise the grass is to be mown twice and the cuttings disposed off site.

Spot treatment of weeds is required where and when directed by the Engineer, including during the Period of Maintenance. This work is not covered in the MMHW.

Minor changes have been made mainly with reference to Appendices rather than to the Contract and with the removal from the SHW to the NG of rules governing spot treatment of weeds.

> An example of a completed Appendix 6/9 is included in Annex A

Box 36

619 • Earthwork Environmental Bunds

There is a general change in description from 'noise' to 'environment', presumably to emphasise the concern for the environment in the vicinity of motorway schemes. With this in mind there is likely to be an increased emphasis in using earth bunds to reduce noise and visual impact.

Appendix 6/9 should stipulate the locations, material requirements and whether the bunds are to be topsoiled, seeded or turfed. For measurement itemisation see Box 37.

Earthworks *environmental* bunds may be reinforced or anchored earth structures, normal or strengthened embankments or landscaped areas. If constructed as a landscaped area it should not be referred to as an earthworks *environmental bund* to avoid confusion.

If there is a requirement for early completion this should be specified in Appendix 6/9 and Sectionalised in the Appendix to the Form of Tender (this will have considerable significance for the programme).

32 — Itemisation

Group	Feature	
I	1	Deposition.
III	5	Landscape areas.
	6	Environmental bunds.

Box 37 MMHW — Deposition Itemisation — Compaction similar

620 • Landscape Areas

If required by Appendix 6/9, method compaction should be used in landscape areas otherwise compaction should be merely sufficient to remove large voids and produce a coherent mass whilst not over-compacting. Such a loose specification could lead to settlement and whilst this would not normally be critical for a landscape area, the measurement rules require the Contractor to allow for this as any fill, deposition and compaction required is not measured (MMHW Series 600 para. 13).

Landscape fill is acceptable fill (Class 4) which may draw in general

Box 38 HA 44/91 — Advice on Landscape Areas

4.37 Landscape areas within the construction site can sometimes be used as a convenient reservoir for acceptable material. They can be increased or decreased as construction proceeds by providing on-site locations for placing excess fill, or by reducing the extent and nature of the landscaping if the amount of acceptance fill material is less than anticipated. Therefore the contouring of a landscape area should not be too rigidly specified to allow the site engineers some latitude in the amounts of material put into it, but of course the basic function of the area must always be of primary importance which is usually to protect or preserve an existing feature. Off-site landscape areas are normally shown in contract documents as part of the site to ensure all provisions apply to them. They usually require planning permission and therefore should be considered permanent features during construction and should not be used to temporarily stockpile material.

12.18 The presence of a landscape area may be beneficial to the earthworks balance by providing a reserve of fill material if there is a shortfall, or a suitable on-site location for placing excess material. Therefore, the design contours should, if possible, leave room for movement and only the general shape or outline of the area provided on the drawings.

Box 39 HA 44/91 — Strengthened Embankments

5.83 The technical literature published by the manufacturers usually provides detailed information regarding the physical properties of their products, but a Designer is advised to acquire evidence that the particular product which the Contractor proposes to use satisfies the design requirements for the reinforcement, and the details of the required evidence should be included in the reinforcement specification. It should be noted that the properties of geotextiles (according to BS 6906) and related materials in isolation may be different to the properties when placed in soil, and care must be excercised that the appropriate material properties are chosen.

11.5 In applications where the polymer reinforcement is taken to the face of the embankment slope, it can be returned back into the fill either at the level of the next reinforcement layer above or at an intermediate level. This will stabilise the face of a steep sided slope susceptible to erosion.

or selected materials and can include those listed within Class U1 unacceptable material in Clause 601. These must be identified in Appendix 6/1 as materials which may be used for landscape fill Class 4. By doing so otherwise unacceptable materials become acceptable by cross-reference to Table 6/1. If this is not done there are major complications with the measurement (see Boxes 37 and 38 for measurement details and advice on using landscape areas as 'reservoirs' of material).

The deposition of landscape fill must await the Engineer's acceptance of the adjoining embankment. This rather strange requirement could lead to complications as the Engineer may not wish to accept part of the Permanent Works prior to the issue of a certificate under clause 48 of the ICE Conditions of Contract. More practically, the landscape area may be filled at the same time as the adjoining embankment keeping 1 metre difference in level. This needs to be specified in Appendix 6/9.

The Seventh Edition Notes for Guidance stipulate that 'environmental bunds' should not be constructed on landscape fill unless special foundations are provided. It would seem that this was intended to read 'environmental barriers' which are described in Series 300.

621 • Strengthened Embankments

Layering of geotextiles and geomeshes in embankments can reduce the future maintenance costs and they enable steeper slopes to be built (see

Box 39). The required properties of geotextiles and geomeshes should be described in Appendix 6/9 together with any construction requirements supplemented by drawings where necessary. A separate feature is provided in the MMHW for strengthened embankments under the itemisation for Deposition of Fill, Imported Fill and Compaction.

622 • Earthworks for Reinforced Earth and Anchored Earth Structures

This clause deals with the earthworks for these structures but the design would normally be carried out by the Contractor (see notes on Clause 106). Series 2500 also specifies details of the structural elements.

Excavation must comply with Clause 604 'Excavation for Foundations' and deposition, compaction, etc. is to be in a line parallel to the facing and the elements are to be placed on the compacted fill.

Types of fill and drainage layers should be specified in Appendix 6/1 and identified on drawings together with the maximum height fill can be placed above the wall (see Box 40). *As these are structures that the Contractor will normally be designing, care should be taken to avoid over-specifying the requirements and thus limiting his choice (see notes on Clause 106).*

Fill for reinforced earth structures should be either Class 6I, 6J, 7B, 7C or 7D subject to any additional requirements stipulated in Appendix 6/1. If Class 7B PFA is used, non-metallic elements shall be used with stainless steel fasteners. Fill to anchored earth structures should be 6I or 6J.

Drainage layers should be 6H, or for use with 7B material should be BS882 fine aggregate and when in contact with metal components should comply with Table 6/3 (*the properties of the fill in Table 6/3 have been considerably changed from the Sixth Edition*). Type B filter drain material should only be used in horizontal drainage layers and vertical layers of drainage should be brought up at the same rate as the adjoining material without mixing or contaminating.

Reinforcing and anchor elements should be kept as free as possible from damage or displacement — 'programme of filling shall be arranged so that no machines or vehicles run on the reinforcing or anchor elements'. It may be wise to repeat this constraint in Appendix 1/13. Plant within 2 m of the back of the facing is restricted and fill beyond the 2 m zone may be raised in thicker layers providing the difference in level is not greater than 300 mm. Retained fill at the rear of the structure is to be maintained at the same level as reinforced earth or anchored earth. This will affect the Contractor's programme. Any shoring to the natural slope or and existing earthwork is to be removed progressively.

Box 40 HA 44/91 — Soil Structures

Structures

11.2 Reinforced and anchored earth techniques are used in the construction of retaining walls and bridge abutments. The term 'reinforced earth' describes a type of soil structure and does not refer to the product of any particular organisation. These types of soil structure comprise a mass of fill which is reinforced by tensile elements and retained by facings which are attached to those elements.

Reinforcement Properties

11.9 Since the design life of soil structures may be as long as 120 years, the durability of the reinforcement and its resistance to degradation is very important and must be considered by the Designer. Where metallic components are used a sacrificial thickness of the metal is usually necessary and a corrosion allowance for metallic components exposed to various environments is given in Table 4 of BE 3/78 (Revised 1987). These allowances should be increased by the appropriate amount in areas where the metallic reinforcement overlaps.

Seperate itemisation is provided in the MMHW for Deposition of Fill, Imported Fill and Compaction.

623 • Earthworks for Corrugated Steel Buried Structures

The words 'of span exceeding 900 mm' have been removed from the clause title.

These structures will normally be designed by the Contractor under the requirements of Clause 106. Measurement of the structure will be on the basis of the Designated Outline and this should include the associated earthworks. Structural requirements in respect of the structure are contained in Series 2500 of the SHW.

If material adjacent to the structure has a low constrained soil modulus or *is of a corrosivity classification determined in accordance with Departmental Standard BD 12/88 at which corrosion of metal components could occur*, the extent of any additional width of excavation should be shown on the drawings (see also Box 41).

There is a new requirement in respect of overlying fill — 6Q or 7H. Method compaction is to be used and the drawings should show the extent of selected fill materials for embankments over the structure. Additional compaction and testing requirements are to be stated in Appendix 6/1.

The lower bedding 6K and surround material 6M should be end-product compacted with layers not exceeding 150 mm. The lower bedding has to be shaped during compaction to match the steel profile. The upper bedding 6L is deposited in a uniform layer to fill the corrugations and is uncompacted (but included in quantities for Compaction — see Box 24).

Sub-clause 6 requires the surround material to be compacted uniformly on either side of the structure. If this wording is meant to convey similar requirements to those clearly stated in SHW Clause 610.4 it does not succeed. It can be interpreted as meaning compacted uniformly to the same degree with no limit on differences in height. *It is unfortunate that the Seventh Edition is as poorly worded as the Sixth in this respect.*

Surround material 6M under the structure is to be compacted to the satisfaction of the Engineer with no other criteria for acceptance. This will require careful supervision and can lead to doubts as to whether the Contractor's price can cover such imprecise requirements.

The plant used for compaction is restricted within 1 m of the side of the structure and above the crown. Sub-clause 10 stipulates that 'only that compaction plant described in Sub-clause 8 above shall be used in the vicinity' unless surround material is 1 m above the crown. *This unaltered wording leads to difficulties in interpretation when compared with the seemingly similar requirements of Sub-clause 8 (these restrictions are presumably meant to be covered in item coverage 33(f), 45(f) and 52(f) — see Box 42).*

Sub-clause 9 places restrictions on changes in shape of the structure during filling, compaction, pavement construction and all other traffic movements. This implies careful monitoring of dimensions and straightness to check compliance. *There are changes in allowable tolerances — the Seventh Edition allows 25 mm for both longitudinal straightness and rotational diplacement compared with 10 mm for both in the Sixth Edition.* Whilst the Clause states that movement shall not exceed these limits there is no indication as to what the position is if they are, nevertheless, exceeded.

Fill to Corrugated Steel Buried Structures

11.11 Where it is intended to construct corrugated steel buried structures in partial or total trench condition, tests must be carried out to measure the corrosivity of the insitu material as well as for the proposed imported bedding and surround material. If these tests have not been carried out for any reason, a provisional sum should be detailed in the Contract Bill of Quantities in case it is later found that the soil is aggressive. This provisional sum can be based either on sacrificial thickness of steel for the structure, or for the provision of additional surround material either side of the structure and above it.

Box 41 HA 44/91 —Corrosivity of material

Deposition of Fill — para. 33 — Item Coverage

(f) complying with the particular requirements and constraints with regard to soil stabilisation, reinforced earth structures, strengthened embankments, anchored earth structures, corrugated steel buried structures and the like;

Box 42 MMHW Item Coverage — Common to Deposition, Imported Fill and Compaction

Measurement rules for Ground Anchorages

123 **The establishment of ground anchorage plant** shall be measured once only to each separate location of ground anchorages on the Site. Any additional establishment of ground anchorage plant to suit the Contractor's method of working shall not be measured.

127 **The measurement of ground anchorages** shall be for the complete anchorage assembly and shall be the length from the bottom of the fixed anchorage to the bearing face.

131 **The measurement of waterproofing of boreholes** shall be the total length of waterproofing operation instructed by the Engineer.

Box 43 MMHW — Ground Anchorages

Normally the Contractor would have designed the structure but the Engineer accepts it after his check — such movement could conceivably be due to poor design or problems of workmanship. The Sixth Edition was identically worded and it is unfortunate that the Seventh Edition has not clarified this area.

No material shall be placed by tipping within 2 m of the structure. *This restriction is presumably covered in item coverage 33(f) (see Box 42).*

624 • Ground Anchorages

The Contractor should normally design ground anchorages for the permanent works and this clause requires him to do so. The Engineer should specify proof loading and design calculations to check the Contractor's design. The requirement for the Contractor to design should be set out in Appendix 1/11 and the design requirements in Appendix 6/10. This should also include installation and construction requirements.

For the rules of measurement see Box 43, although it will often not be possible to be precise in detailing the quantities when the Contractor is designing the structure. This aspect does not seem to have been adequately covered in the measurement rules for 'Structural Elements'.

Temporary works anchorages will be permitted only with the approval of the Engineer.

625 • Crib Walling

The specific requirements of the Sixth Edition are omitted as the Contractor is required to design crib walling under Clause 106. The design requirements should be set out in Appendix 6/10 which should include an outline Approval in Principle form where the retained height exceeds 1.5 m and the Overseeing Department must be consulted when the design requirements are being formulated. There are no requirements in the SHW in respect of crib walling apart from this clause. Series 2500 Special Structures does not mention them.

Where the Contractor is responsible for design the Designated Outlines should allow for the full range of alternative systems. Otherwise the measurement rules are set out in the MMHW paragraphs 113 to 116.

626 • Gabions

Retaining walls of gabions will often be designed by the Contractor under SHW Clause 106 and measured by Designated Outlines although there are no standards or approval systems. However there are measurement rules in the MMHW (paras 109–112) for situations where the Engineer designs them and for the Contractor in preparing his schedule of quantities.

The Sixth Edition clause is changed in only minor ways — 2.2 mm core diameter compared with 2.0 mm (SHW 626.3(i)) and dark green or black PVC coating (SHW 626.4).

The requirements for filling, etc. are rather imprecise — 'sufficiently

Box 44 HA 44/91 — Voided Ground

> 9.17 Underground voids are either man-made or of natural orgin. Natural cavities are usually the result of the flow of water through soluble rocks. If they still form part of an underground watercourse they may present special difficulties, since any measures to fill the cavities could interfere with the ground water flow and could cause problems upstream or downstream. Therefore the consequences of such a filling should be thoroughly investigated beforehand. In some cases cavities are disguised by being covered with detritus, such as is sometimes found in swallow holes in chalk and magnesian limestone, and are only discovered during construction. It is therefore sensible to allow for this when the scheme passes through these materials, and a detailed specification of the treatment required for such natural voids included in the Specification. One suggested method of dealing with swallow holes is to fill them with a granular material, sealed over by a layer of clay or other relatively impermeable material such as lean-mix concrete.

filled', 'allowance for consolidation', 'minimise distortion', 'where appropriate be maintained square', 'maintain tightness of mesh'.

Mesh is to be as described in Appendix 6/10 either to 3(i) or 3(ii) and all wire must be galvanised and with PVC covering. Mesh openings and grading are to be described in Appendix 6/10 with the maximum size of the fill material not more than two-thirds the minimum dimension of the gabion (or 200 mm) and the minimum size not less than the mesh opening (unless otherwise stated in Appendix 6/10).

The sample appendix no longer contains the note regarding the 'ratio of size of fill to mesh openings if different from 626.5' even though the SHW clause still requires this information.

Mechanical filling is allowed only if the Engineer is satisfied that the results are equivalent to filling by hand. The use of the word 'equivalent' makes this a potentially difficult area with differences of opinion on what constitutes equivalent and with a significant affect on pricing.

627 • Swallow Holes and other Naturally Occurring Cavities

There are no significant changes to this clause from the Sixth Edition.

Full requirements for excavating, filling, compacting and capping must be given in Appendix 6/11. Other earthworks clauses do not cover these, so precise requirements need to be specified (see Box 44).

The clearance of rubbish in open swallow holes is only to be carried out if it will not endanger operatives. This rather imprecise wording makes this aspect difficult to price which is perhaps why it seems to be missing from the item coverage (see Box 45 for measurement rules).

628 • Disused Mine Workings

There is no significant change to the SHW but an additional note is given in Sample Appendix 6/11 regarding requirements for inspection, monitoring, clearing, flushing, filling, etc.

Box 45 MMHW Filling and Caps to Swallow Hole and Mine Workings

> 118 The measurement of filling to mine working, well, swallow hole and the like shall be calculated from the tonnage of material certified by the Engineer, being only that material, included on delivery tickets, which is incorporated in the Permanent Works in the locations and to the extent and thickness required by the Contract. Material in excess of the requirements of the Contract and material used for any other purpose shall not be included within the certified tonnage.
>
> 119 The measurement of caps to mine working, well, swallow hole and the like shall be the volume of concrete forming the caps.

Appendix 6/11 should specify the location of the disused workings together with methods of identifying and inspecting them and filling, grouting, flushing and capping. It may be necessary to specify the equipment to avoid noise and dust nuisance during grouting operations, e.g. sound reduction requirements (see also HA 44/91 para. 9.19). It is important to specify the extent of work anticipated so that the Contractor can programme and price it on a known basis (see Box 45 for MMHW measurement rules). The method of testing compliance should be detailed — in grouting this is particularly important and the position regarding retests should be clearly stated.

629 • Instrumentation and Monitoring

There is no significant change to the SHW but the Notes for Guidance no longer refer to the need to extend the Method of Measurement if the Contractor is to monitor the instruments. The MMHW gives reasonably extensive coverage including establishment of plant; boring holes; instrumentation; erection, servicing and dismantling instrument huts or cabinets; and monitoring equipment. Full details and requirements including location and extent, instrument types, housings, calibration, protection, electric power and reporting should be specified in Appendix 6/12.

Where instrumentation is vital to the control of construction, an additional specification clause should be included in the Contract stating that if the instrumentation is damaged or the data is affected or discontinued then no further work will be allowed until the equipment is repaired. Care should be taken to ensure that the responsibility in respect of payment and delay are clearly identified.

Instrumentation for blasting and ground improvement should not be included in this appendix but in Appendix 6/3 or 6/13.

630 • Ground Improvement

There is virtually no change from the Sixth Edition.

Only one system of dynamic compaction should be used on any one Contract, either end-product or method compaction (see Box 46 for advice and Box 47 for measurement rules).

Dynamic compaction must be complete before any permanent works or Statutory Undertaker's work is commenced within that part of the site. The precise area must be defined in Appendix 6/13 as this could be a considerable constraint on the Contractor's programme. It should therefore

Box 46 HA 44/91 Ground Improvement and Dynamic Compaction

Ground Treatment

9.13 Ground improvement details should be provided in SHW Appendix 6/13. Earthworks details for excavation and replacement etc should be detailed on the drawings and in SHW Appendix 6/3. The Designer should be aware of the Specification for Ground Treatment and Notes for Guidance published by ICE in 1987.

Dynamic Compaction

9.22 Whereas DC has been used extensively in Continental countries for the treatment of soil on roads and other civil engineering sites, experience in Great Britain is more limited. The Engineer is, therefore, often obliged to follow the advice provided by the specialist firms which carry out this process which, in many cases, will result in an end-product specification being followed rather than a method specification. In other cases the choice between method and end-product specification will depend on the type of material to be compacted, the conditions on site and the experience of either the Engineer or the Contractor or both with the materials and methods used and the quantity of testing required for end-product specification.

98 The establishment of dynamic compaction plant shall be measured once only to each separate location of dynamic compaction on the Site. Any additional establishment of dynamic compaction plant to suit the Contractor's method of working shall not be measured.

102 The measurement of dynamic compaction shall be the sum of the distances through which the pounder is required to fall. The distance for each drop shall be the vertical measurement from the underside of the pounder immediately prior to release, to the level of the ground beneath the pounder immediately prior to the first drop at that point.

103 The measurement of dynamic compaction plant standing time shall be for the period or periods of standing time ordered by the Engineer. Periods of less than half an hour shall not be measured. Any other standing time due to the Contractor's method of working, necessitated by the process of ground improvement provided for in the Contract or other than that ordered by the Engineer shall not be measured.

104 The measurement of granular blanket shall be the tonnage of material certified by the Engineer, being only that material included on delivery tickets which is incorporated within the Permanent Works in the locations to the extent and thicknesses stated in the Contract or ordered by the Engineer.

Box 47 MMHW Ground Improvement Measurement

be noted in Appendix 1/13 although this is not one of the constraints listed in the sample appendix.

631 • Earthworks Material Tests

Appendix 6/1 should state whether the Contractor or the Engineer is to be responsible for testing and details should be given in Appendix 1/5. *Not surprisingly the Notes for Guidance no longer contain the note stating the 'number of earthworks tests and their frequency will not normally be forseeable at the tender stage' as the designer now has to predict it. Where the Engineer is to carry out the tests, details of sampling should be shown in Appendix 1/6.*

NG 631.1 recommends allowing rapid evaluation tests and these 'acceptable tests' should be described in Appendix 6/1. These will presumably then be tests for compliance but the Engineer is required to monitor them using the basic test periodically.

Unusual tests should be carried out in a commercial testing laboratory *accredited by NAMAS (see notes on Clause 105).*

632 • Determination of Moisture Condition Value (MCV) of Earthworks Materials

There are minor changes in references to British Standards.

Appendix 6/1 should specify whether the determination of the MCV/mc relationship should be carried out. The rapid assessment procedure for acceptability may be used when permitted by the Engineer. Unlike SHW Clause 631 there is no requirement for this to be shown in Appendix 6/1.

633 • Determination of Undrained Shear Strength of Remoulded Cohesive Material

This clause has been considerably reworded with increased reference to British Standards and omission of detailed requirements of the Sixth Edition 633.1 (i)–(v).

Appendix 6/1 should specify where undrained shear strengths should be determined by triaxial compression tests. If 'other tests' are specified in Appendix 6/1 it should be made clear whether these are compliance tests. *The Engineer may also permit or agree to such tests but if these were not included in Appendix 1/5 then the Contractor will not have allowed for their cost.*

636 • Determination of Effective Angle of Internal Friction and Effective Cohesion of Earthworks Materials

There are many small changes in this clause with the omission of detail and some sub-clauses compared with the Sixth Edition.

637 • Determination of Resistivity to Assess Corrosivity of *Soil, Rock or* Earthworks Materials

The details of each test are no longer included in the SHW clause and there are new requirements in respect of orientation and depths compared to the Sixth Edition. The Notes for Guidance no longer contains the note on the orientation of the electrodes presumably because of the additional requirements of SHW 637.4.

The test is to be either in-situ or in the laboratory as stipulated in Appendix 6/1 and Appendix 1/5. Where laboratory tests are specified, the appendix should state which of the three types described in BS 1377 should be used.

SHW Sub-clause 3 states that details of the tests should be given to the Engineer but with no mention of approval. Notes for Guidance Clause 637.1 sets out points which the Engineer should take into account in considering the Contractor's proposition which seems to point to a possible conflict. *The Seventh Edition has not resolved this point..*

638 • Determination of Redox Potential (E) to Assess Corrosivity of Earthworks Materials *for Reinforced Earth and Anchored Earth Structures*

This clause has been considerably reworded with references to BS 1377 and substantial omission of detail previously set out in the Sixth Edition Sub-clauses 638.3/4/5/6(iii)–(iv) and 638.7. Reference is made to Appendix 6/1 in respect of testing but surprisingly the Notes for Guidance do not refer to Appendix 1/5. Appendix 6/1 is to state if laboratory tests are acceptable, otherwise they are to be in-situ.

Details of area to be tested, etc. are to be given to the Engineer but there is no mention of approval in SHW Sub-clause 2. Notes for Guidance Clause 638.1 lists points which the Engineer should take into account in considering the 'Contractor's proposals'. *As in Clause 637 there seems to be a conflict between the documents unresolved by the Seventh Edition.*

If pH is outside the range 5.5 to 9.5 and is known to remain so for the life of the structure, redox potential need not be measured.

639 • Determination of Coefficient of Friction and Adhesion between Fill and Reinforcing Elements or Anchor Elements for Reinforced Earth and Anchored Earth Structures

There are only minor changes to the SHW compared with the Sixth Edition.

Tests should be carried out on each type of reinforcing element and each fill material. Surprisingly the Notes for Guidance do not refer to Appendix 1/5. There is no criterion for acceptability in the clause — it should be specified in Appendix 6/1.

Although SHW Sub-clause 639.5 allows for tests on anchor elements the test has not yet been developed and such a test should not be specified until it has.

642 • *Determination of the Constrained Soil Modulus of Earthworks Materials*

This is a new clause setting out the procedure for determining the constrained soil modulus by means of a plate loading test.

Appendix 6/1 should indicate whether such a test is to be carried out. The requirement is for three tests on each side of the structure to be carried out unless otherwise stated in Appendix 6/1 or required by the Engineer. Two procedures are described — one for compacted granular fill and one for 'material existing on site'. If the Engineer is not satisfied with the results of the first load cycle the procedure is repeated at a new location.

There is no reference in the Notes for Guidance to this test being included in Appendix 1/5.

Table NG 0/2 List of Sub-clauses which permit Contract-specific requirements to be included in the Contract instead of the national ones stated, e.g. Sub-clauses state '. . . unless otherwise described in Appendix -/-'

Earthworks	
601.2 (ii)	Composition of unacceptable material Class U1.
602.12	Battering.
603.3	Halting of excavation.
605.1 (v)	Maximum vehicle capacity for trafficking Class 3 material.
605.1 (x)	Treatment of Class 3 material at end of working day.
607.2 (vii)	Arrangements for installing instruments off the Site.
608.4	Construction of fills.
610.1 (iv)	Fill above structural concrete foundations.
612.1	Compaction of fills.
612.10 (ii)	Areas of fill requiring extra compaction.
613.3	Capping materials and layers.
613.8	Longitudinal gradient, crossfall and surface tolerances of sub-formation.
614.2	Cement type for stabilisation.
618.4 (i)	Application of fertiliser and seed by hydraulic mulch.
618.5 (ii)	Harrowing of cutting slopes.
618.13	Seed mixture.
619.1	Deposition and compaction to environmental bunds.
626.3	Gabion mesh.
626.5	Minimum size of fill material to gabions.
633.1	Undrained shear strength tests.
636.1	Shear box or triaxial test methods.
642.1	Selection of location for constrained soil modulus test.

Table NG 0/3 List of Sub-clauses which require the Contractor to submit information to the Engineer.

Note. Information that the Contractor may submit when seeking the Engineer's approval is not listed in this table.

Earthworks	
603.4	Pre-split blasting — submit details of methods, etc.
607.2 (ii)	Blasting — give written notice of each blasting event.
607.2 (vii)	Blasting — submit details of proposed instrumentation and report results daily.
608.7	Surcharge material — submit details.
609.2	Geotextiles — provide evidence of durability.
612.6	Compaction methods — submit proposals for compaction trials.
612.12 (i)	End-product compaction — submit maximum dry densities and optimum moisture contents.
612.12 (ii)	End-product compaction — submit density/moisture content graph.
615.4	Lime stabilisation — submit analysis reports.
617.3	Formation protection proposals — submit proposals where formation is within 300 mm of existing ground level.
618.13	Grass seed — provide certificates of germination and purity.
629.2	Instrumentation — supply results of monitoring.
637.3	In-situ resistivity tests — submit details.
637.6	In-situ resistivity tests — submit proposals for testing at depth.
638.3	In-situ redox potential tests — submit details.
638.4	In-situ redox potential tests — give notice of testing arrangements.

Fourth Edition of the Method of Measurement for Highway Works

In commenting on the Fourth Edition of the Method of Measurement the problem arises of what to leave out in the knowledge that the relevant items in the MMHW will be cross-referenced with the commentaries on the SHW Series. It was considered that the basic underlying principles as set out in Chapters I to III were essential and that notwithstanding that items are cross-referenced throughout the notes, certain topics required special attention. These form the subjects for 'Selected Topics'.

Introduction

The Fourth Edition of the MMHW and the Fourth Edition of the Library of Standard Item Descriptions have been based on the Seventh Edition of the Specification for Highway Works and the Fourth Edition of the Highway Construction Details. In addition they are intended to be used with the Fifth Edition of the ICE Conditions of Contract. Prior to this, an addendum to the Third Edition of the MMHW was published in October 1991. This addendum included changes which had become necessary, both technically and administratively, as a result of four years use of the Third Edition MMHW and the Sixth Edition SHW.

The Fourth Edition MMHW includes most but not all of the changes introduced by the addendum with additional changes which have been deemed necessary in order to avoid creating barriers to trade within the European community. The Fourth Edition MMHW has three new organisational concepts. The first is that the Sections are now renamed 'Series' which is in line with the annotation of the SHW. The second is that there are now Notes for Guidance. This should be welcome to many users as this useful aid was not available with the Third Edition. The third concept is that the Fourth Edition will be revised on an annual basis. This is a prudent move on behalf of the DTp. and will remove the inevitability of amendments having to wait for a new edition to be published possibly some four years later.

The MMHW Notes for Guidance are not part of the Contract Documents but it is envisaged by the DTp. that should a situation arise where the MMHW is brought into question then the Notes for Guidance may be used as supporting evidence.

Basic Principles

The MMHW is a document that has nothing to do with the way a Contractor actually carries out the work. The item coverage in the Bill of Quantities is not a substitute for the SHW, HDC or Drawings. The MMHW is purely a mechanism through which a Contractor can tender and be paid for work done.

The item coverage within the MMHW must be predictable, so that the Contractor knows the items of work to be covered by the rates and prices he inserts against the items in the Bill of Quantities. Some item coverage lists are extensive but this does not mean that they are all-inclusive. If a particular item listed is not supported by drawings detailing this work or is not reasonably implied then the Contractor cannot and should not price this item.

Consider the situation where the item coverage for hard material includes cutting through reinforced concrete (MMHW Series 600: para. 23(d)). A tenderer who prices for hard material will not price for this item unless the contract documents at the time of tender imply that material of this nature is to be found. Similarly, some activities (fencing, safety fencing, kerbing, etc.) include an item in the item coverage for excavation in hard material. This does not imply that if hard material is encountered the Contractor has always allowed for it in his rates. He is required to do so only when the 'nature and extent of the works' as set out or reasonably implied from the contract documents indicates he will have to deal with such material. The Notes for Guidance to the MMHW confirm this view and recommend to the Engineer that information about the presence of hard material including buried roads and the like should be included in the Contract.

Words such as 'as directed by the Engineer' found in the SHW create uncertainty as it is impossible at tender stage for a Contractor to know what the Engineer will actually direct once the contract is underway. The Contractor's estimate can only be based on what can be reasonably implied from the Contract and if the Engineer directs work in excess of this then the Contractor will usually be entitled to additional payments under the Conditions of Contract.

In an attempt to make the MMHW less protracted, the item coverage often refers to item coverage set out in one or more other Series and, as such, the complete item coverage embodies all these references. This cross-reference is necessary and was used in the Third Edition MMHW but, due to poor drafting, created ambiguity in certain Sections. The wording of the Fourth Edition MMHW has been amended to remove these uncertainties, notably the 600 Series item coverage for Deposition of Fill and Imported Fill apparently including compaction.

The general obligations of the Contractor set out in, amongst other documents, the Conditions of Contract are not separately itemised in the Bill of Quantities but instead are covered in the Preambles to Bill of Quantities set out in the MMHW. The testing requirements of the Contract as scheduled in Appendix 1/5 would be deemed to be included in these general obligations set out in para. 2 (vii) of the Preambles.

If changes are made to the SHW or the HCD this will necessitate a review of the item coverage to ensure accurate agreement but the item coverage should not be extended to include items which are not specified or shown in the HCD. Changes to the SHW should not be proposed through the Drawings although revised Drawings may be required for revisions to the SHW.

Definitions

(i) 'Hard Material' means one of the following:

(i) *material so designated in the Preambles to Bill of Quantities*;

(ii) material which requires the use of blasting, breakers or splitters for its removal but excluding individual masses less than 0.20 cubic metres;

Box 1 MMHW Chapter 1 Definitions

MMHW Chapter 1

Definitions

Paragraph 1(i), the definition of Hard Material, has been amended, initially by the October 1991 addendum, to be those deposits so designated in the Preambles to the Bill of Quantities and material in excess of 0.2 m^3 that require removal by blasting, breakers or splitters (see Box 1 and Selected Topics).

Paragraph* 1*(l)* *Type of Pavement. *Definitions have been included for the pavement types that are offered as design options in the Contract (see Box 2).*

Paragraph* 1*(m)* *Designated Outline. *The definition of Designated Outline as required by SHW Clause 106 (Design of Permanent Works by the Contractor) is given in Box 3, and these are shown on a Drawing as an outline enclosing a structure to be designed by the Contractor (see Selected Topics).*

MMHW Chapter 2

General Principles

Paragraph 1(b) Method of Measurement. *The MMHW is based on the SHW and the HCD published as Volume 1 and Volume 3 of the MCD and on the principle that full details of construction requirements are provided in the Contract.*

This principle is of fundamental importance to the MMHW and additions or amendments to the SHW or to the HCD which are not covered adequately by the MMHW will require an amendment to the MMHW. Provision is made in Chapter III Preparation of Bill of Quantities for this to be accommodated.

Paragraph 2(a) Bill of Quantities. *An amendment introduced by the October 1991 addendum and maintained in the Fourth Edition MMHW removed the HCD as being a reference requirement for the nature and extent of the work to be performed under the contract. However, the drawings from the HCD are brought into the contract by reference through the List of Drawings Included in the Contract as detailed in Appendix 0/4. The cross-reference to the HCD in the Third Edition MMHW was therefore superfluous.*

Paragraph 2(c)(i) and (ii) Bill of Quantities. *Work required within and below water has to be identified in the Bill of Quantities. Sub-paragraphs 2(c)(i) and (ii) have been amended to include the word 'datum' in brackets as meaning the 'open water level' or 'tide level' as appropriate.*

Definitions

(l) 'Type of Pavement' means one of the following designs of pavement:

(i) flexible;
(ii) flexible composite;
(iii) rigid;
(iv) rigid composite;

Box 2 MMHW Chapter 1 Definitions

MMHW Chapter 3

Preparation of Bills of Quantities

The main changes in this chapter relate to amendments to the Preambles to Bill of Quantities and to the introduction of design options and alternatives and the related amendments to Table 1. The amendments to Table 1 include the requirement for Motorway Communications to be listed as a construction heading. Level 1 Division, sub-para. (iii) Structures has been amended to add Pocket Type Reinforced Brickwork Retaining Walls, but Corrugated Steel Structures are deleted. These are now included in Structures Designed by the Contractor. Level 1 Division has been extended to include:

(iv) Structures Where a Choice of Design is Offered.
(v) Structures designed by the Contractor.

Paragraph 1 Sub-division of Bill of Quantities. *The note relating to Table 1 level 4 sub-headings as being advisory has not been amended. The level 4 sub-headings shown in Table 1 refer to non-tidal water and tidal water. This has created an ambiguity over the requirement to measure separately work to be carried out in tidal or non-tidal conditions as required by the Preambles to Bill of Quantities para. 2(c). The 'Note' should not be considered an alternative to the Preambles requirement.*

Paragraph 4 Special Preliminary. *Special preliminaries should be used at the discretion of the Engineer for temporary works whose operations and costs are unusual in relation to the Works or for operations required in advance or after the main Works.*

Paragraph 5 Alternative Types of Pavement. *This was introduced by the October 1991 addendum and details the requirements for separate Bills to be prepared for Series 600: Earthworks; Series 700: Pavements, of the Main Carriageway, Interchanges and Side Roads for each of the Pavement options the Contractor can price. Provision is to be made for only one of the Bills in Series 600 and Series 700 to be priced and included in the Tender Total. The measurement is to be based on the thinnest pavement permitted for the particular pavement choice.*

For each separate Bill, an Index must be prepared and inserted immediately preceding each Bill. MMHW Table 2 is the index to be inserted

Definitions

(m) 'Designated Outline' means the designated outline shown on the Drawings.

Note. A Designated Outline is shown as enclosing each structure to be designed by the Contractor and each structure for which a choice of designs is offered. The Designated Outline delineates the limits of measurement of work to be included for each structure (with the exception of those works scheduled as not to be included).

Box 3 MMHW Chapter 1 Definitions

as a separate page immediately preceding each set of the separate Bills of Quantities included within Series 600: Earthworks in the Roadworks General Bill and Series 700: Pavements in the Main Carriageway, Interchanges and Side Roads Bills, to cover the alternative Types of Pavement included in the Contract.

Paragraph* 6 *Alternative Types of Safety Fence. *This was introduced by the October 1991 addendum and details the requirements for separate Bills to be prepared for the Series 400: Safety Fences, Safety Barriers and Pedestrian Guardrails for the two options permitted by the Contract, namely wire rope safety fence or tensioned corrugated beam safety fence. Provision is to be made for only one Bill in Series 400 to be priced and included in the tender total. For each Bill an Index must be prepared (MMHW Table 3) and inserted immediately preceding each Bill.*

Paragraph* 7 *Structures Where a Choice of Designs is Offered. *This is new to the Fourth Edition MMHW and details the requirements for Structures Where a Choice of Designs is Offered.*

Where the Contract provides for a Structure to be designed by the Contractor in accordance with SHW Paragraph 106 and listed in Appendix 1/10(B) as an alternative to the Engineer's design, a separate Bill of Quantities is to be provided for each of the two construction procedures offered by the contract. Each Bill of Quantities is to be provided in accordance with the appropriate Series in the MMHW for all the works contained in the Designated Outline.

For Structures designed by the Contractor, the Bill of Quantities is to consist of a single item in accordance with the Series 2500 and priced as a lump sum. The Bill of Quantities for the Structure designed by the Engineer is to be compiled from the appropriate Series in the MMHW.

Provision is to be made for only one Bill of Quantities relating to the form of construction and elected by the Contractor to be priced and included in the Tender Total. Immediately preceding the separate Bill of Quantities an Index (MMHW Table 4) is to be provided detailing the alternative forms of construction permitted by the contract.

Paragraph* 8 *Structures Designed by the Contractor. *This is a new paragraph but retains the principles introduced by the October 1991 addendum relating to buried structures designed by the contractor.*

Where the contract provides **only** *for a Structure to be Designed by the Contractor and constructed in accordance with SHW Paragraph 106 and listed in Appendix 1/10(A), a Bill of Quantities consisting of a single item for all the work within the Designated Outline (except those scheduled not to be included) is to be prepared in accordance with Series 2500. The works scheduled not to be included in the Designated Outline are to be included by the Engineer in other Bills.*

Paragraph* 9 *Preambles to Bill of Quantities. *This is a new paragraph but self explanatory. The Preambles to Bill of Quantities are always to be included as a preamble to the Bill of Quantities but may be added to if necessary and amendments to the MMHW are to follow paragraph 17 of the Preambles.*

Preambles to Bill of Quantities

The preambles have been added to and amended. They total 17 paragraphs; paragraph 17 relates to amendments to the MMHW.

Paragraph* 1 *General Directions. *This has been amended to include the Fourth Edition MMHW as the definitive document. The procedure for the annual amendments to the MMHW are that the publication date of each page of the MMHW is given in the Schedule of Pages and Relevant Publication Dates. This schedule and the completed Preambles to the Bill of Quantities must be reproduced unaltered and bound into the Bill of Quantities.*

Paragraph 2(vi) *has been reworded by the October 1991 addendum to exclude specific reference to existing services and supplies although this could be implied by the new words of 'any element of the works'.*

Paragraph 2(vii). *This has not changed but its importance cannot be overstressed.*

The NG MMHW for Series 100 paragraph 5 states that responsibility for repairing damage to the highway rests with the Contractor but this should be interpreted in the light of the 'Expected Risks', but see commentaries under SHW 117 and 118.

Paragraph 2(x) *has been amended by the October 1991 addendum to include providing test certificates.*

Paragraph 2(xi) *has been amended by the October 1991 addendum to include providing certificates of conformity for QA schemes.*

Paragraph 3(ii) Measurement, *has been amended by the October 1991 addendum and further amended by the Fourth Edition MMHW. The Paragraph sets out that the measurement of pavement shall be based on the thinnest pavement construction permitted by the contract. Where the contract provides for the Contractor to select safety fencing, pavement or buried structure then the measurement of all work in each area shall also be based on the thinnest pavement construction permitted by the contract.*

Paragraph 5 Alternative Specified Materials, Designs and Options within Types of Pavement. *This has been reworded by the October 1991 addendum. The rates and prices the Contractor inserts in the Bill of Quantities are deemed to cover any of the permitted alternative materials or designs the Contractor has elected to use. This applies to any inherent permitted options within the Bills of Series 600 and Series 700. In all cases the rates and prices shall include any adjustment of work content, rates, costs incurred by the Contractor as a result of his choice. With measurement being based on the thinnest alternative pavement construction it follows that actual construction could be different from what is measured. Tenderers must be aware of the possible consequences of inappropriate rates.*

Paragraph 9 Work Within and Below Non-tidal Open Water or Tidal Water. *This has been amended by the Fouth Edition MMHW. The 'datum' stated in the contract shall be used for the measurement of this work 'Subject to and without prejudice to the Condition of Contract'. The Contractor has to allow in his rates for carrying out work of this kind that is separately measured as such. Actual water levels could be different from the datum depending on seasons and state of tide.*

The Notes for Guidance to the MMHW on Chapters I, II, and III of the MMHW, paragraph 3, Non-tidal and Tidal Water, offers guidance on the measurement of work within these conditions. For measurement in non-tidal water the NG states that the datum indicated on the drawing should be as close as possible to the actual level at the time of tender. During the contract period this level may change considerably especially during the winter periods. In contrast to this, the NG states that the datum for tidal water should be based on the highest astronomical spring tide. Should the datum for non-tidal water not be based on a similar supposition i.e. on the highest flood level?

No separate measurement is provided for, and datums may or may not be stated for items of structures designed by the Contractor under Series 2500, therefore the Contractor must allow in his rates for work of this nature.

Paragraph 12 Site Limitations and Constraints. *This is a new paragraph. The Contractor has to allow for constraints and limitations on the use of the site but only to the extent of Appendix 1/17 as required by SHW Clause 107.*

Paragraph 13 Hard Material. *This forms one of the Selected Topics.*

Paragraph 14 Equivalent Products and Materials. *This is a new*

Structures Designed by the Contractor

The items for structures designed by the Contractor shall be in accordance with the Preambles to Bill of Quantities General Directions include for:

(a) design;
(b) certificates;
(c) preparation and submission of priced schedule of quantities;
(d) provision of data and drawings;
(e) awaiting approvals;
(f) resubmissions and modifications;
(g) amendment to the Works;
(h) obtaining aesthetic approval;
(i) everything necessary for the completion of the Works within the Designated Outlines, as shown in the relevant item coverage in the Chapters and Series of the Method of Measurement, with the exception of those works scheduled as not to be included;
(j) completing an Approval in Principle Form and submitting to the Engineer for acceptance.

Box 4 Series 2500: Special Structures. Item coverage

paragraph and was introduced by the October 1991 addendum. Where a Contactor offers an alternative product or material in accordance with SHW Clause 104, then the Contractor's rates and prices shall be deemed to include for **all** *obligations and costs associated with the incorporation of this material or product into the works. This is to include design, provision of data and drawings, certificates, awaiting approvals, resubmissions, modifications and amendments to the Works. The measurement of the works affected by the inclusion of the product or material is to be based on the original tender and not on the works as amended by the equivalent product or materials. The actual costs of submissions will be difficult to estimate at tender stage. The Conditions of Contract Clause 8 has been amended by the addition of a sub-clause 8B(3) to the effect that the costs of the Engineer's examinations and checks for the Contractor's design or proposals shall be borne by the Employer. The NG SHW Clause NG 104 gives details of equivalence procedures to be carried out by the Engineer upon submission of a Contractor's proposal.*

Paragraph* 15 *Permanent Works Designed by the Contractor. *This is a new paragraph introduced by the October 1991 addendum. The rates and prices the Contractor inserts in the Bill of Quantities shall include for all obligations and costs associated with the incorporation of the Contractor's design into the works. This shall include all design, data, drawings, certificates, approvals and modifications and amendments to the Works. Amended Conditions of Contract Clause 8 apply.*

Paragraph* 16 *Structures Designed by the Contractor. *This is a new paragraph. For each lump sum priced Bill of Quantities for Structures designed by the Contractor in accordance with SHW Clause 106 and Series 2500, the Contractor shall prepare a priced schedule of quantities in accordance with the appropriate MMHW Series and including the items*

MMHW — Preambles to Bills of Quantities

Hard Material

13 *For the purposes of the Contract the following are designated as Hard Material in accordance with Chapter 1 Definitions, paragraph 1 (i):*

*(a) . . . * strata;*
(b) those deposits designated by limits shown on the Drawings;
(c) existing pavements, footways, paved areas (but excluding unbound materials) and foundations in masses in excess of 0.20 cubic metres.

Box 5 Preambles to Bill of Quantities. Hard Material

in Series 2500 (see Box 4). The extended total shall be equal to the lump sum price. The priced schedule shall only include the parts of the works within the Designated Outline.

The priced schedule shall be submitted to the Engineer who shall accept it within 7 working days of the date the Engineer approves in writing the Contractor's design. The schedule is for the Engineer's acceptance, not approval, and the grounds for non-acceptance would appear to have little or no basis.

The schedule shall be used for the valuation of monthly statements and for the valuation of ordered variations under Clause 52 of the Conditions of Contract. The Engineer may also prepare a Bill of Quantities in a similar way where a specific designed option is included for such a structure.

The measurement of the Works affected by the Contractor's design shall be based on the Tender documents and not on the Contractor's incorporated design. Earthworks within the Designated Outline shall not be included in the earthworks schedule.

Selected Topics

Hard Material

Excavation in Hard Material occurs in many of the Series and its occurrence and extent must be adequately described in the Contract so that the Contractor's tender can allow for it in the rates.

The Engineer must make available to the tenderers information relating to Hard Material in the form of existing buried pavement structures which may not be apparent to a tenderer complying with Clause 11 of the Conditions of Contract.

Hard Material is warranted as being measured as extra over normal excavation due to the increased costs associated with its removal. *In bulk earthworks materials which in the opinion of the Engineer can reasonably be removed by using conventional plant and taking into account location and extent should not be designated as Hard Material.*

If material is designated as Hard Material in the tender documents then irrespective of how hard the material actually is, it shall be measured as Hard Material. If the nature and extent of the material found during excavation is different in character from that which was envisaged at the time of tender then whilst not affecting the measurement, this will not preclude the Contractor from pursuing a claim for additional payment under the appropriate clause of the Conditions of Contract.

Problems arise when the extent of the designated strata is not clear, e.g. where the upper and lower surfaces become indistinguishable from the adjacent layers. If Hard Material cannot be accurately defined in the Contract by reference to the geological strata then it should be designated by limits. The limits can be difficult to decide and must therefore rely on the professional judgement of the designer.

The Preamble to the Bill of Quantities (see Box 5) sets out the three methods of designation for Hard Material for measurement purposes. The designer selects one of these alternatives to describe any hard material that he anticipates will be encountered and bases his quantities on this. Bound materials in existing pavements and the like will always be Hard Material but unbound materials within the pavement, footway, paved area

Box 6 Notes for Guidance to SHW — Sample Appendix 6/1

Sample Appendix 6/1

4 *Any requirement for processing to render unacceptable material Class U1 acceptable, cross-referring to Drawings where necessary, for each type of material to be processed and class of material to be produced [Wherever possible the means of processing should be left to the Contractor] [601.4].*

Box 7 MMHW Series 600: Itemisation for Processing of Unacceptable Material Class U1

Itemisation Group	Feature
I	*1 Processing of unacceptable material Class U1.*
II	*1 Different locations.*
III	*1 Into different classes of acceptable material*

or foundation would be excluded. The Contractor then prices accordingly, making an assessment of the likely difficulty of excavation. Significant increases or decreases in quantity would be subject to Conditions of Contract Clause 56(2) which, because of the wording of the hard material item coverage, could affect rates other than the extra-over for Hard Material.

It is essential that the Engineer ensures that only one method of designation is used for any one particular material. A material designated as Hard Material is not subject to re-classification. Conversely, the fact that a material similar to that designated as Hard Material is found elsewhere does not indicate that it will also be classified as Hard Material.

Where Hard Material is designated by strata alone then the total quantity excavated from that strata is subject to admeasurement . Where Hard Material is designated by limits shown on the Drawings, then that volume is measured and paid for as Hard Material. In both cases the measurement is not affected by whether the material requires special plant — once designated as Hard Material it remains as Hard Material for measurement purposes.

The Engineer must ensure that the quantities in the Bill of Quantities are consistent with the information made available to the tenderers.

Processing Materials

A new Clause 601.4 Series 600 allows the designer to specify that U1 unacceptable material shall be rendered acceptable by processing. The details are to be described in Appendix 6/1 (see Box 6). The actual means of processing should be left to the Contractor. Material which is required to be processed as such shall be measured in accordance with Processing of Unacceptable Material Class U1 in MMHW Series 600: paras 24–28. The itemisation for Processing of Unacceptable Material Class U1 is as Box 7. It is important that the Engineer states the class or classes of material that the processed material should comply with.

If the Contractor wishes to render unacceptable material acceptable for use in the Contract, as opposed to when the Engineer has specified this, then the above measurement rules will not apply. In this case measurement shall be as though the unacceptable material had been disposed of and the material rendered as acceptable by processing as being imported.

Designated Outlines *(see also Series 100 Notes)*

Where the Contract requires a structure to be Contractor-designed in accordance with SHW 106 and listed in Appendix 1/10(A) or 1/10(B) then a Designated Outline is required on the Drawings. (This is not required for structural elements described in Appendix 1/11).

Each outline should be large enough to include all the possible options the Contractor may put forward including any special backfill requirements, etc. The Designated Outline delineates the limits of measurement of work to be included for such structures and as such should be clearly shown and fully enclosing. Features which pass through or into the outline, such

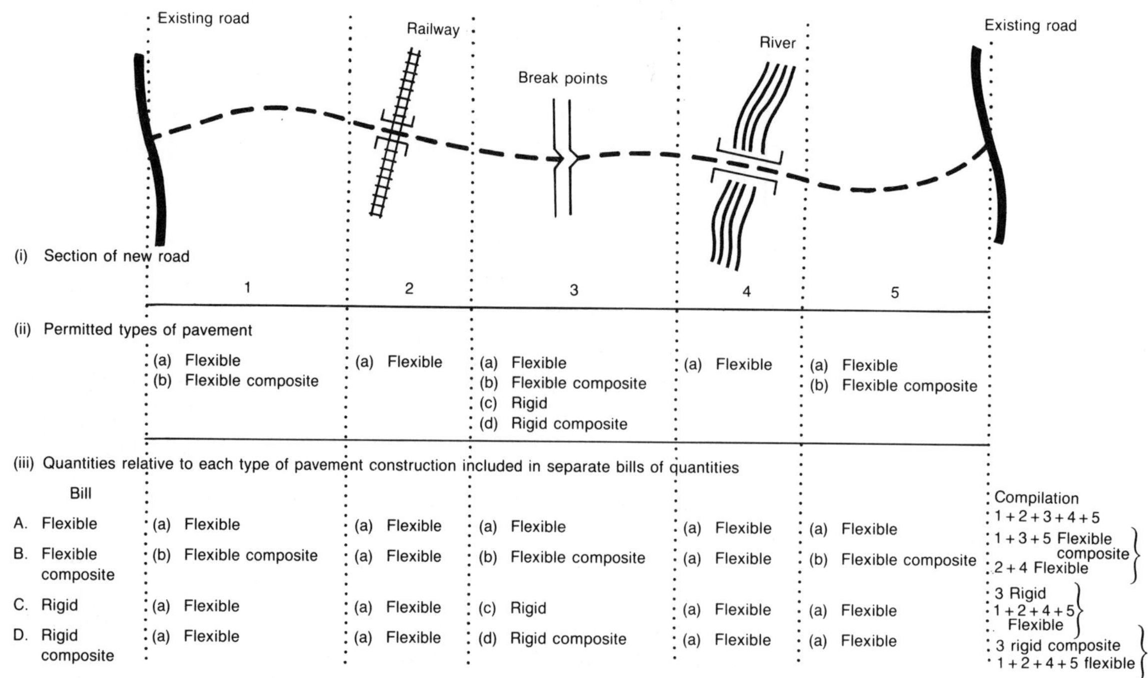

Fig. 1. Example of how bills of quantities relative to differing types of pavement construction for a main carriageway are compiled

Box 8 MCD Modified ICE Clause 1(1)(r)

> ICE 5th Edition Definitions. Clause 1(1)(r)
>
> *'temporary diversion for traffic' means:*
>
> *(i) a temporary carriageway onto which vehicular traffic is diverted from a highway;*
>
> *(ii) a temporary footpath or bridleway onto which pedestrian or equestrian traffic is diverted from a highway;*
>
> *(iii) a combination of (i) and (ii) or a temporary carriageway as in (i) with an associated footway and/or way for the use of animals and equestrian traffic; or*
>
> *(iv) a temporary private means of access onto which traffic is diverted from a private means of access;*
>
> *but in all cases shall not include a central reserve crossover constructed to permit contraflow traffic on an existing carriageway.*

as pavements, kerbing, safety fencing, service ducts, headwalls, etc., should be clearly scheduled to be excluded from the lump but included in other Bills. The measurement of works outside the Designated Outline but which have been affected by the Contractor's design shall be measured as the Tender documents and not as amended by the Contractor's design. The Contractor is required to allow for the cost of these changes under paragraph 16 of the Preambles to Bill of Quantities.

Contraflows *(see also Series 100 Notes)*

The Conditions of Contract Clause 1 has been amended (See Box 8) to include a definition for Temporary Diversions for Traffic (SHW Clause 118) and specifically excludes the case of a central reserve crossover to permit contraflow. BS 6100 has been amended to include a definition for contraflow (see Box 9).

The measurement for the above is included in the MMHW Series 100: paras 22–25 (see Box 10 for itemisation). The NG MMHW recommends that the Contraflow item is always included when traffic management is required, thus allowing for the Contractor's proposals.

> BS 6100 — Contraflow
>
> Temporary flow of two traffic streams in opposite directions routed on one side of a dual carriageway road.

Box 9 BS 6100 No 241 7131

Alternative Pavements

With the many permitted design options, accurate compilation of the Bills

Itemisation

24 Separate items shall be provided for traffic safety and management in accordance with Chapter II paragraphs 3 and 4 and the following:

Group	Feature
I	1 Traffic safety and management.
	2 Taking measures for or construction, maintenance, removal of contraflow arrangements.

Box 10 MMHW Series 100: Contraflows. Itemisation.

Proofloading.

Itemisation

42 Separate items shall be provided for proof loading of piles in accordance with Chapter II paragraphs 3 and 4 and the following:

Box 11 MMHW Series 1600: Proofloading.

of Quantities is important. The NG MMHW illustrates by example the compilation of a Bill of Quantities for a road project which comprises several sections of the road each with a different choice of design (see Figure 1).

The measurement of each Bill shall be based on the thinnest construction permitted for each type of pavement. For each Type of Pavement a separate Earthworks Bill of Quantities is required for each of the pavement types. The tenderer only prices and extends the Bill appropriate to the chosen pavement option. The position is similar for pavements measured under Series 700 and provision is to be made for only one Bill of Quantities in Series 600 and in Series 700 to be priced and included in the tender total.

Testing and Proofloading *(see also Series 100 Notes)*

The Overseeing Department in their Model Contract Document amend Conditions of Contract Clause 36(3) to require testing to be at the Contractor's own cost 'provided that the test in question is particularised in the Specification in sufficient detail to enable the Contractor to have priced or allowed for the same in his Tender'. The tests shall be abstracted from Table NG 1/1 and scheduled in Appendix 1/5 if they are to carried out by the Contractor.

There is a certain lack of clarity in the case of proof of design testing which NG SHW Clause NG 105.2 states should be separately itemised. It is not clear that this is the case with the amended Clause 36 although the original ICE clause would have required it.

Proofloading is an itemisation in MMHW Series 1600: Piling and Diaphragm Walling (see Box 7), but there are no other examples of proof of design itemisation. The October 1991 Addendum introduced items for Proof Demonstration of Capping Construction which have not been continued in the Fourth Edition and the tensile loading of anchorages in bolt holes in Series 2200 is not itemised and is to be scheduled in Appendix 1/5.

Communications

This is a complete new series to the MMHW and has been based on the Overseeing Department's TCC Division Motorway Communications Manual Volume III Document which had previously been incorporated under contract documents. The National Motorway Communication System (NMCS) covers telephones, matrix signals and their power supplies but does not extend to TV monitoring systems to motorways, other roads or to Temporary Diversions for Traffic (SHW Clause 118).

Should there be a requirement for TV monitoring for temporary works then this work should be a separate Bill item in the MMHW Series 100, rather than extending the item coverage for traffic diversions. If TV monitoring is to be for permanent works then new Bill items should be raised following the principles of Series 1500 (see NG MMHW).

Curved Formwork
Formwork shall be measured as curved of whatever radius if the final finish to the concrete surface is a curved surface. This infers that the curve could be made up of series of straights. If curved formwork is used during construction but with the object of producing a flat finish once the concrete has been placed then the formwork shall not be measured as curved. Formwork for curved falls and cambers shall be measured as curved.

Accommodation Works
There is no SHW Series for Accommodation Works. However, the MMHW Series 2700: Accommodation Works states that the measurement of this work shall be in accordance with other series within the MMHW as appropriate. If the exact nature of work for private and publicly owned services and supplies is not known then it would be appropriate to insert provisional quantities or provide a Provisional Sum preferably in that order.

Main Changes to the Third Edition MMHW

Chapter I

1(i)(i)	*The word 'Strata' replaced by 'materials'.*
1(l)	*Type of pavement added.*
1(m)	*Designated Outlines added.*

Chapter II

2(c)	*The word 'datum' added.*

Chapter III

5	*Alternative Types of Pavement added.*
6	*Alternative Types of Safety Fence added.*
7	*Alternative Structures added.*
8	*Structures Designed by the Contractor added.*

Table 1 contains amendments to match changes to other parts of the MMHW.

Tables 2, 3 and 4 added.

Preambles

1	*New paragraph inserted.*
3(ii)	*New sub-paragraph.*
5	*Amendments regarding alternatives and options inserted.*
9	*New sub- paragraph.*
12, 13, 14, 15, 16	*New paragraphs.*

Notes at end of chapter amended

Series 100

3	*Note added.*
13(g)	*Item coverage added.*
21(ii	
23	*[illegible] contr[illegible] f[illegible]ws added.*
24	

25	*Item coverage amended and re-lettered.*
26	*Item coverage for contraflows added.*
28	*Measurement instructions added.*
29	*Features amended.*
30	*Item coverage amended.*
31	*Item coverage amended.*
32	*Item coverage amended.*

Series 200

An introductory note covers lowering carriageway levels.

5(h)	*amended.*
5(j) *5(k)* *5(l)*	*added.*
7 to 10	*Additional materials provided for.*

Series 300

Hedges deleted and Pedestrian guardrails moved to Series 400.

Noise Barriers renamed Environmental Barriers.

17	*Item coverage amended.*

Series 400

Safety barriers and pedestrian guardrails added.

2	*Wire rope definition added.*

Safety Fences renamed Beam Safety Fences.

3(ii)	*Provision for additional items.*
4 to 7	*Amendments to terms.*
8 Grp I	*Features added.*
8 Grp II	*Curvature limits amended.*
8 Grp III *8 Grp IV*	*New Groups*
9 to 13	*Item coverage amended.*
14	*Item coverage added.*
15 to 57	*Provisions for additional items.*
18 to 22	*Item coverage added.*
24	*Group I amended.*
27	*Group I amended.*
28	*Item coverage amended.*
29 to 37	*Wire rope provision added.*
38 to 42	*Concrete safety barriers added.*

Pedestrian Guardrails added.

45	*Group III added.*

Series 500

7	*Note added.*
10	*Contractor's Design paragraph added.*
12	*Additions and amendments to item coverage.*
13	*Provisions for additional items.*
14 to 16	*Directions amended*
17	*Group I and II features added.*
18 *19*	*Additions and* *amendments to item coverage.*
20 to 22	*Narrow Filter Drains added.*

23 *31* *32*	*Additions and amendments to item coverage.*

Channels and Combined Drainage and Kerb Blocks now measured in Series 1100.

36	*Group added.*
37 *50* *57*	*Additions and amendments to item coverage.*

Series 600

1(c)	*Amended.*
11	*Additional paragraph for Designated Outlines*
15(a)	*Amended.*
15 (h) & (i)	*New sub-paragraphs*
16	*Features added.*
18	*Amendments to item coverage.*
21	*Amended.*
22	*Features added.*
23	*Amendments to item coverage.*
24 to 28	*Processing paragraph added.*
32	*Feature added.*
33	*Amendment to item coverage.*
36	*Amended.*
37	*New sub-paragraph.*
41	*Amended.*
44	*New Feature added.*
45	*Amendments to item coverage.*
51	*New Feature added.*
52	*Amended item coverage.*
55	*Amended.*
62(a)	*Amended.*
63	*New Feature added.*
65 to 66	*Amended Heading.*
66 to 72	*Amended.*
80 *81*	*Item coverage* *added.*
81 to 82	*Amended Heading.*
84	*Item coverage added.*
86	*Amended.*
87	*Feature amended.*
90	*Amended.*
92 *112*	*Item coverage* *added.*
118	*Amended.*
121	*Item coverage added.*
128	*Feature amended.*
129	*Amended item coverage.*
145 *146*	*Item coverage* *added.*
162	*Item coverage amended.*
167 to 170	*New paragraphs*

Series 700

4 *8*	*Feature* *amended.*
9	*Item coverage amended.*

16	*Amended description.*
19	*Item coverage amended.*
20 to 24	*New paragraphs.*

Heading Amended

25 to 27	*Terms and item coverage amended.*
32	*Item coverage amended.*
34	*Measurement rules amended.*
36	*Item coverage amended*
37 to 41	*New paragraphs.*

Series 1100

Heading Amended

1 *2* *3*	*Scope increased.*
4	*Item coverage amended*

Heading Amended

5 *6* *7*	*Scope increased.*
8	*Item coverage amended.*

Heading Amended

9 *10*	*Scope* *increased.*
11 *12* *21*	*Item coverage* *amended.*
27 to 30	*New paragraphs.*

Series 1200

2	*New paragraph.*
4 *5*	*Item coverage* *amended.*
11	*Amended.*
15	*Features amended.*
18	*Item coverage amended.*
21	*Features amended.*
22	*Amended description.*

Heading Amended

23 to 25	*Amended description.*
27	*Amended.*
29 *30* *34* *35*	*Item coverage amendments.*
39 to 41	*New paragraphs.*

Series 1300

2	*New paragraphs.*
4	*Item coverage amended.*

Series 1400

1 to 4	*New paragraphs.*

Paragraph	Change
6	*Amended.*
7	*Amended.*
8	*Features amended*
9 }	*Item coverage*
10 }	*amended.*
11	*New paragraphs.*
13	*New paragraphs.*
15 }	
18 }	*Item coverage amended.*
19 }	
20 to 23	*New paragraphs.*

Series 1500

New series.

Series 1600

Paragraph	Change
14 }	*Item coverage*
16 }	*amended.*
19	*Amended.*
25 }	*New*
26 }	*paragraphs.*
27	*Amended Groups and Features.*
61	*Item coverage amended.*
62	*Amended.*
65	*Item coverage amended.*
66 }	*New*
67 }	*paragraphs.*
68	*Amended Feature.*

Series 1700

Paragraph	Change
2	*New sub-paragraph.*
3	*Amended Features.*
4 }	*Item coverage*
9 }	*amended.*
15	*Item coverage amended.*
22	*Amended.*
24	*Features and Groups amended.*

Heading Amended

Paragraph	Change
27(i)	*Description added to.*
28	*Amended.*
29	*Feature amended.*
33	*Amended.*
36	*Item coverage amended.*

Series 1800

Paragraph	Change
2	*Amended.*
6 }	*Item*
9 }	*coverage*
18 }	*amended.*

Series 1900

Paragraph	Change
4	*Item coverage amended.*

Series 2000

Paragraph	Change
4	*Item coverage amended.*
5 to 8	*New paragraphs.*

Series 2100

3	*Group added.*
4	*Item coverage amended.*

Series 2200

1	*Description added to.*
2	*Amended.*
3	*New paragraphs.*
4	*Amended.*
5	*Item coverage amended.*
6	*New paragraphs.*

Series 2300

5, *10*	*Item coverage amended.*

Series 2400

4, *8*	*Item coverage amended.*
9 to 12	*New paragraphs.*

Series 2500

New series.

Series 2600

No change.

Annex A: Extracts from Earthworks Advice Note

A** HERNCASTER BYPASS

APPENDIX 6/1

CLASSIFICATION AND ACCEPTABILITY CRITERIA

CLASS	GENERAL MATERIAL DESCRIPTION	TYPICAL USE	PERMITTED CONSTITUENTS	MATERIAL PROPERTIES				COMPACTION REQUIREMENTS
				PROPERTY	TEST	LIMITS		
						LOWER	UPPER	
1B	Uniformly Graded	General Fill Drawings Nos 305/6/1B-5	Any materials other than Class 3	grading	BS 1377: Part 2	Tab 6/2	Tab 6/2	Tab 6/4 Method 3
				uniformity coeff	ratio of D_{60} to D_{10}	-	10	
				optimum mc	BS 1377: Part 4 (2.5kg rammer method)	-	-	
				mc	BS 1377: Part 2	Opt mc - 2%	Opt mc + 2%	
1C	Coarse Granular Material	General Fill Drawing Nos 305/6/1B-5	Any materials other than Class 3	grading	BS 1377: Part 2	Tab 6/2	Tab 6/2	Tab 6/4 Method 5
				uniformity coeff	ratio of D_{60} to D_{10}	5	-	
				10% fines value	Clause 635	50 kN	-	
2B	Dry Cohesive Material	General Fill Drawing Nos 305/6/1B-5	Any materials other than Class 3	grading	BS 1377: Part 2	Tab 6/2	Tab 6/2	Tab 6/4 Method 2
				MCV	Clause 632	13	16	
2D	Silty Cohesive	General Fill Drawing Nos 305/6/1B-5	Any materials other than Class 3	grading	BS 1377: Part 2	Tab 6/2	Tab 6/2	Tab 6/4 Method 3
				MCV	Clause 632	8	15	
4	Various	Landscape Area Fill	Any materials	grading	BS 1377: Part 2 Passing 500 mm Passing 63 mm	 - 10%	 100% 100%	Clause 620.2
				MCV	Clause 632	6	18	
6F1	Selected Granular Material (Fine)	Capping Drawing Nos: 305/6/3A-5 305/7/2-5B	Any materials other than unburnt colliery spoil and argillaceous rock	grading	BS 1377: Part 2	Tab 6/2	Tab 6/2	Tab 6/4 Method 6
				optimum mc	BS 1377: Part 4 (Vibrating hammer method)	-	-	
				mc	BS 1377: Part 2	Opt mc - 2%	Opt mc	

A** HERNCASTER BYPASS

APPENDIX 6/1 (Cont'd)

CLASSIFICATION AND ACCEPTABILITY CRITERIA

CLASS	GENERAL MATERIAL DESCRIPTION	TYPICAL USE	PERMITTED CONSTITUENTS	MATERIAL PROPERTIES				COMPACTION REQUIREMENTS
				PROPERTY	TEST	LIMITS		
						LOWER	UPPER	
6J	Selected Uniformly Graded Granular	Fill to Reinforced Earth Drawing No 305/10/1 305/10/2B	See Table 6/1	grading	BS 1377: Part 2	Tab 6/2	Tab 6/2	Tab 6/4 Method 3
				uniformity	ratio of D_{60} to D_{10}	5	10	
				mc	BS 1377: Part 2	17%	21%	
				effective c' effective ϕ'	Clause 632	50 kN/m^2 25°	- -	
				coeff of friction adhesion	Clause 639	0.6 50 kN/m^2	-	
6P	Selected Uniformly Graded Granular	Fill to Structures Drawing Nos: 305/9/1/5 305/9/2/4 305/9/3/6 305/9/4/5&6 305/9/6/4	See Table 6/1	grading	BS 1377: Part 2	Tab 6/2	Tab 6/2	95% maximum dry density of BS 1377: Part 4 (Vibrating Hammer Method)
				uniformity coeff	ratio of D_{60} to D_{10}	5	-	
				10% fines value	Clause 635	30 kN	-	
				undrained shear c ϕ	Clause 633	50 kN/m^2 20°	- -	
				effective shear c' ϕ'	Clause 636	40 kN/m^2 25°	-	
				coefficient of permeability	Clause 640	5 x 10^4 m/sec	- -	
				mc	BS 1377: Part 2	17%	21%	
7E	Selected Cohesive Material	For Stabilisaitcn With Lime to Form a Capping (Class 9D)	Any material or combination of materials, other than unburnt colliery spoil	grading	BS 1377: Part 2	Tab 6/2	Tab 6/2	Not applicable
				MCV	Clause 632	7	13	
				plasticity index	BS 1377: Part 2	10	-	
				organic matter	BS 1377: Part 3	-	2%	
				total sulphate content	BS 1377: Part 3	-	1%	

CLASS	GENERAL MATERIAL DESCRIPTION	TYPICAL USE	PERMITTED CONSTITUENTS	MATERIAL PROPERTIES				COMPACTION REQUIREMENTS
				PROPERTY	TEST	LIMITS		
						LOWER	UPPER	
7F	Selected Silty Cohesive Material	For Stabilisation With Cement To Form A Capping (Class 9B)	Any material, or combination of materials, other than chalk, unburnt colliery spoil and argillaceous rock	grading	BS 1377: Part 2	Tab 6/2	Tab 6/2	Not applicable
				uniformity coefficient	ratio of D_{60} to D_{10}	5	-	
				MCV	Clause 632	8	15	
				Liquid Limit	BS 1377: Part 2	-	45	
				Plasticity Index	BS 1377: Part 2	-	20	
				organic matter	BS 1377: Part 3	-	2%	
				total sulphate content	BS 1377: Part 3	-	1%	
7G	Selected Conditioned Pulverised Fuel Ash. Cohesive Material	For Stabilisation With Cement To Form A Capping (Class 9C)	Conditioned material direct from power station dust-collection system and to which a controlled quantity of water has been added	mc	BS 1377: Part 2	15%	35%	Not applicable
				total sulphate content	BS 1377: Part 3	-	1%	
9B	Cement Stabised Silty Cohesive Material	Capping	Class 7F with addition of cement according to Clause 614	pulverisation	BS 1924: Part 2	30%	-	Tab 6/4 Method 7
				MCV immediately before compaciton	Clause 632	8	12	
				bearing ratio	BS 1924: Part 2	15%	-	
9C	Cement Stabilised Conditioned Pulverised Fuel Ash. Cohesive Material	Capping	Class 7G with addition of cement according to Clause 614	pulverisation	BS 1924: Part 2	60%	-	End Product 95% of maximum dry denisty of BS 1924 (2.5 kg rammer method)
				bearing ratio	BS 1924: Part 2	15%	-	
				mc	BS 1924: Part 1	To enable compaction to Clause 612		
9D	Lime Stabilised Cohesive Material	Capping	Class 7E with addition of lime according to Clause 615	pulverisation	BS 1924: Part 2	30%	-	Tab 6/4 Method 7
				MCV immediately before compaction	Clause 632	8	12	
				bearing ratio	BS 1924: Part 2	15%	-	

APPENDIX 6/1 (Cont'd)

Classification

The classification and confirmation of acceptability of the earthworks materials shall be carried out by the Contractor at the point of excavation for on-site materials, and at the point of deposition for imported materials. Trial pit locations for classification purposes shall be agreed with the Engineer in advance. If in the opinion of the Engineer the material has altered its classification or become unacceptable for whatever reason, he may require the Contractor to repeat the classification and acceptability tests given in Table 6/1 and this Appendix. The rate of testing required shall be as stated in Additional Clause 656 AR, Appendix 0/1.

The Contractor shall submit two copies of all test results to the Engineer within one working day of the completion of the test. The copies shall be signed by the Contractor's responsible engineer/ technician.

Class 3 Material

No material is designated as Class 3.

GroundWater Lowering - Cut No 3 (sub-Clause 602.17)

After stripping topsoil, but before commencing any bulk excavation between Ch 3570 and Ch 4090 in cut No 3, the Contractor shall construct a deep temporary cut-off filter drain in the position and to the invert levels as shown on Drawing No 305/5/17A. This drain shall be maintained to allow ground water to outfall freely into the stream at Ch 3570 (South side) until the permanent road drainage through Cut No 3 is completed. A full description of these works is given in Additional Clause 652 AR, Appendix 0/1.

Permeability Test (Clause 640)

The permeability test referred to in Clause 640 shall be the constant head or constant hydraulic gradient permeability test as described in BS 1377: Part 5 and Part 6 for vertical permeability. The permeability box test described in HA 41/90 is to be used for horizontal permeability.

APPENDIX 6/3

EARTHWORKS REQUIREMENTS

Drawings

305/6/1B
/2
/3A
/4
/5

Generalearthworks drawings showing quantities, cut and fill numbering and chainages, selected fill locations, location of unacceptable material, simplified GIborehole information and soil strata.

Blasting

Blasting is a permitted alternative within the confines of Cut No 4 (Ch 5200 to Ch 6500) for the excavation of the Lower Lincolnshire Limestone only.

Pre-split blasting shall also be employed if the limestone is to be excavated by bulk blasting. Details of the diameter, dip, azimuth, spacing and depth of the pre-split drill holes together with the minimum panel size and face lift are given in Additional Clause 643 AR, Appendix 0/1. Other specification requirements for pre-split blasting including the trial are given in Additional Clause 644 AR, Appendix 0/1.

(Note: For details of pre-split blasting procedures, specifications etc see TRRL LR 1094; pre-split blasting for highway rock excavation, and also TRRL SR 817; A device for measuring drill rod and drill hole orientations).

Blasting of any kind shall only take place between the hours of 0930-1130 and 1430-1630 Mondays to Fridays, and 1000-1200 Saturdays. No blasting shall be carried out on Sundays or Bank Holidays.

Cutting Faces

In Cut No 2 (Ch 2900-3340) South side, the verge drains within the silty cohesive material shall be excavated in such a manner that only 20 m of drain trench over 1 m deep may be open at any one time.

Excavation of further lengths of trench may not commence until a sufficient length of any existing trench has been backfilled and fully compacted so that the 20 m length of open trench is not exceeded.

All cut-off drains and ditches alongside cuttings and

embankments shall be completed and outfalls provided prior to the commencement of any adjacent earthworks excavation or filling operations.

The faces of cuttings requiring attention prior to topsoiling shall be treated in accordance with sub-Clause 603.7(i)(a), with the exception of Cut No 4 where the cut faces in limestone shall not be topsoiled but treated in accordance with sub-Clause 603.6(i).

Watercourses

The existing stream at Ch 2010-2040 North side shall be realigned and regraded to the line and level as shown on Drawing No: 305/5/6. Concrete lining, 150 mm thick, as shown on Drawing No: HBS/5/17B shall be placed on the invert and that side slope adjacent to the embankment from Ch 2005-2050. The redundant stream bed shall be treated in accordance with sub-Clause 606.4 and backfilled with Class 1B material. Drainage pipes along the redundant stream bed shall be as detailed in Drawing No: 305/5/6.

Embankment Construction

The Contractor shall not allow fills of more than 2 m height to remain at side slopes of 1:2.5 or steeper for more than 48 hours before trimming back to the design slope, and in any event the side slope shall not be steeper than 1:2 at any stage of construction.

No surcharging of embankments is required, and the Contractor shall not stockpile material on fill areas to a height greater than that of the finished embankment. See also Additional Clause 649 AR, Appendix 0/1.

The minimum thickness of capping material when used as a weather protection layer shall be 450 mm and the minimum thickness of sub-base when used as a weather protection layer shall be 150 mm. See sub-Clause 608.7.

Compaction

The additional compaction for the top 600 mm below sub-formation is not required for the farm access tracks at Ch 8052 and Ch 4990. Elsewhere sub-Clause 612.10(ii) shall apply.

Nuclear Density/Moisture gauges may be used for the measurement of field dry density. In Classes 6P and 9C materials, BS 1377: Part 4 (vibrating hammer method) and BS 1924: Part 2 (2.5 kg rammer method) respectively shall also be used as a basis for compliance.

APPENDIX 6/5

GEOTEXTILESEPARATORS

Drawings

305/6/1B
/6

Location

A geotextile shall be used under the fill material for the full width of Bank No 1. It shall be laid directly on top of the existing ground surface after topsoil has been stripped between Ch 1500 and Ch 1800. The geotextile shall be manufactured from synthetic fibres and shall have a life expectancy of 40 years.

Design Criteria (sub-Clause 609.4)

The geotextile shall sustain a tensile load of not less than 10 kN/m and have a minimum axial strain of 20% at failure. It shall also, have a minimum water flow rate through it at right angles to its principal plane, in either direction, of 50 litres/m^2/S. See Substitute Clause 659 SR, Appendix 0/1.

Sampling

Samples for testing in accordance with Clause 609 shall be taken at the rate of 1 set of samples per 400 m^2. A set of samples shall consist of that minimum number of test pieces sufficient to carry out all the tests required in Clause 609.

Installation

The geotextile shall be laid from rolls in a longitudinal direction along the line of the bypass, and jointing shall be by lapping only. Physical jointing is not permitted. The lap width shall be 500 mm minimum at any location. See Substitute Clause 659 SR, Appendix 0/1.

APPENDIX 6/6

FILL TO STRUCTURES

Drawings

305/9/1/5 Bridge No 1 - Fill to Abutments
/2/4A No 2 - Fill to Abutments and Piers
/3/6 No 3 - Fill to Abutments and Piers
/4/6B No 4 - Fill to Abutments
/5/5 No 5 - Fill to Abutments
/5/6 No 5 - Fill to Piers

Location and extent of selected granular fill Class 6P material.

The Contractor is required to show that his proposed material is stable when compacted and trimmed to a slope of 1 vertical to 1.5 horizontal as described in sub-Clause 610.6.

APPENDIX 6/7

SUB-FORMATION AND CAPPING

Drawings

305/7/2	305/6/3A	HBS/7/1
/3B	/4	to
/4	/5	HBS/7/12
/5B		

Pavement drawings, earthworks drawings and standard detail drawings showing the extent, locations, widths and thickness of Capping materials.

Sub-formation shall have the same shaping requirement as formation as shown in Drawing Nos HBS/7/1 to HBS/7/12.

Where formation is formed in the Lower Lincolnshire Limestone in Cut No 4 between Ch 5200 and Ch 6500 the material shall, in accordance with sub-Clause 616.4, be either:

1) excavated to a depth of 500 mm and the material crushed to give a maximum particle size of 500 mm or

2) where the surface is tabular, it shall be regulated where necessary with a cement bound material Class CBM2 as specified in Clause 1037.

APPENDIX 6/8

TOPSOILING AND SEEDING

Drawings

305/11/1A
/2B
/3A

Topsoil

No imported topsoil Class 5B is required.

As stated in sub-Clause 618.3, topsoil shall not be excavated from stockpiles which have been exposed to accumulative rainfall of 150 mm over the preceding 28 days measured at the main site offices. For details of protection of topsoil stockpiles see Additional Clause 655 AR, Appendix 0/1.

The areas to be grassed and the treatment each area shall receive are shown on Drawing Nos: 305/11/1A and 2B.

Topsoil depths to be deposited in Treatments I and II are 225 mm on slopes of 10° except those areas shown on Drawing No: 305/11/3A as being areas of tree planting where the topsoil depth shall be 300 mm.

All turfing on slopes of gradient of 1:3 or steeper shall be pegged as described in Additional Clause 656 AR, Appendix 0/1.

The hydraulic mulch seeding used in Treatment III shall contain glass fibre as a retaining agent.

Areas of grass which require to be mown 3 times in the vicinity of Bridge No 5, East abutment are detailed in Drawing No: 305/11/3A.

The locations, access points and approximate contouring of permanent topsoil storage areas are shown on Drawing No: 305/11/3A.

APPENDIX 6/9

LANDSCAPE AREAS ETC

Drawings

305/10/1
/2B
/3
/4
/5

Location of noise bunds and landscape areas, cross-sections and contours, details of dense planting areas.

Environmental Bunds

Environmental Bund No 1 shall be constructed using a strengthened embankment to Clause 621. Environmental Bund No 2 shall be constructed as a normal embankment to Clause 619 except that Appendix 6/1 shall apply instead of Table 6/1, using Class 1B material.

Environmental Bund No 1 is detailed on Drawing Nos 305/10/1 and 2B using a geosynthetic reinforcement material at the vertical spacings as shown. The reinforcing material shall comply with the requirements of Additional Clause 660 AR, Appendix 0/1 and shall be laid and jointed as detailed in Additional Clause 661 AR, Appendix 0/1. The fill material shall be Class 6J material as detailed in Appendix 6/1, laid and compacted in accordance with Clauses 608 and 612.

The reinforcing material shall comply with the following additional requirements:-

1. Tensile strength of 20 kN/m at a maximum strain of 10% according to BS 6906: Part 1 (1987) except that the strain rate shall be set at 2% per minute ± 0.4% per minute.

2. Water flow of 30 litres/m^2/s minimum under a constant 100 mm head of water according to BS 6906: Part 3 (1989).

3. Minimum Tear Strength of 400 N when tested in accordance with ASTM D-4533-85 (Geotextilesonly).

4. Minimum Puncture Resistance of 2kN when tested in accordance with BS 6906: Part 4 (1989).

5. Be unaffected by acid/alkali or biological attack.

6. Contain an UV inhibitor if left exposed for more than 7 days in cumulative time.

7. Have a design life of 40 years.

Landscape Areas

Locations of Landscape Area Nos 1-4 together with details of access points and contours are given in Drawing Nos 305/10/3, 4 and 5.

Compaction of Landscape Area Nos 1, 2 and 4 shall be in accordance with sub-Clause 620.2. Compaction of Landscape Area No 3 shall be in accordance with Table 6/4 Method 4.

Construction of Landscape Area No 3 may be carried out at the same time as the adjacent Fill No 4 subject to the conditions described in sub-Clause 620.4.

All landscape areas shall be topsoiled and seeded in accordance with Treatment I, Clause 618. The minimum depth of topsoil shall be 300 mm throughout.

EXCAVATION												LOCATION
ACCEPTABLE						UNACCEPTABLE				Total Excavation other than Class 5A	E.O. Hard Material	
5A	3		Other than Class 3 and Class 5A		Total Acceptable other than Class 5A	U1		U2				
	Above Earthworks Outline	Below Earthworks Outline	Above Earthworks Outline	Below Earthworks Outline		Above Earthworks Outline	Below Earthworks Outline	Above Earthworks Outline	Below Earthworks Outline			
1	2	3	4	5	6	7	8	9	10	11	12	
25	0	0	104	0	104	130	0	0	0	234	130	MAIN CARRIAGEWAY 100-160
135	0	0	19	90	109	28	50	0	150	337	28	MAIN CARRIAGEWAY 160-400
659	245	303	3171	252	3971	0	50	0	0	4021	447	MAIN CARRIAGEWAY 400-590
85	0	0	56	57	113	0	50	0	0	163	0	MAIN CARRIAGEWAY 590-820
3907	402	431	35835	747	37415	11719	241	0	0	49375	7035	MAIN CARRIAGEWAY 820-1520
97	0	0	94	64	158	0	50	0	0	208	0	MAIN CARRIAGEWAY 1520-1660
425	0	0	2133	171	2304	0	50	0	0	2354	0	MAIN CARRIAGEWAY 1660-1975
904	0	0	3028	0	3028	309	50	0	0	3387	309	MAIN CARRIAGEWAY 1975-2143
1511	0	0	6367	822	7189	0	50	0	0	7239	0	MAIN CARRIAGEWAY 2143-2840
38	0	0	0	26	26	0	50	0	0	76	0	MAIN CARRIAGEWAY 2840-3400
1981	0	0	13919	835	14754	0	50	0	0	14804	2015	MAIN CARRIAGEWAY 3400-3534
5232	0	0	5071	0	5071	250	88	0	0	5409	336	INTERCHANGE 3534-3680
715	0	0	0	0	0	649	625	0	0	1274	0	MAIN CARRIAGEWAY 3680-4010
102	0	0	0	0	0	379	13	0	0	392	202	MAIN CARRIAGEWAY 4010-4150
369	0	0	8023	0	8023	92	50	0	0	8165	92	SIDE ROADS-SOUTH LANE
274	0	0	0	182	182	130	14	0	0	326	14	SIDE ROADS-STRAIGHT STREET
459	0	0	0	306	306	0	0	0	0	306	0	AMENITY BUND-SOUTH OF STRAIGHT STREET
1924	0	0	0	1283	1283	0	0	0	0	1283	0	LANDSCAPING-SOUTH OF STRAIGHT STREET
269	0	0	0	180	180	0	0	0	0	180	0	AMENITY BUND-NORTH OF STRAIGHT STREET
834	0	0	0	556	556	0	0	0	0	556	0	LANDSCAPING-NORTH OF STRAIGHT STREET
												WATERCOURSES
0	0	0	0	0	0	0	50	0	0	50	0	NORTH DRAINAGE OUTFALL
45	0	0	0	0	0	252	0	0	0	252	0	SOUTH DRAINAGE OUTFALL-NEW
0	0	0	0	0	0	200	0	0	0	200	0	SOUTH DRAINAGE OUTFALL-ENLARGED
0	0	0	0	0	0	500	0	0	0	500	0	INTERCEPTING DITCHES
	647	734	77820	5571		14638	1531	0	150			Sub Totals
19990	1381		83391		84772	16169		150		101091	10608	ROADWORKS TOTALS

FILL															
GENERAL FILL				LAND-SCAPE	SELECTED GRANULAR							SELECTED COHESIVE			Total Fill Material
				4	6A	6B	6C	6D	6E	6F	6G	7E	7F	7G	
Embankments etc.	Strengthened Embankments	On Sub-Base, Capping etc.	Environmental Bunds	Fill to Landscape Areas	Starter Layer (Below Water)	Starter Layer (Coarse)	Starter Layer	Starter Layer under PFA	Capping for Cement Stabilisation	Capping	Fill to Gabions	Capping for Lime Stabilisation	Capping for Cement Stabilisation	PFA Capping for Cement Stabilisation	
13	14	15	16	17	18	19	20	21	22	23	24	25	26	27	28
82	0	4	0	0	0					0	0	0			86
3927	1150	17	0	0	0					227	0	528			5849
94	0	13	0	0	0					252	0	0			369
10920	0	32	0	0	645					586	0	884			13067
179	0	49	0	0	0					988	0	0			1216
5945	0	14	0	0	0					266	0	402			6627
76	0	8	0	0	0					171	0	0			255
50	0	7	0	0	0					0	0	0			57
145	0	41	0	0	0					822	0	0			1008
435	0	13	0	0	0					240	0	363			1051
128	0	44	0	0	0					835	0	0			1007
892	0	10	0	0	0					0	0	169			1071
2561	0	39	0	0	0					789	0	265			3654
461	0	9	0	0	0					0	0	82			552
185	0	0	0	0	0					0	150	0			335
1492	500	0	0	0	0					0	0	0			1992
0	0	0	6811	0	0					0	0	0			6811
0	0	0	0	24729	0					0	0	0			24729
0	0	0	1959	0	0					0	0	0			1959
0	0	0	0	8705	0					0	0	0			8705
50	0	0	0	0	0					0	0	0			50
0	0	0	0	0	0					0	0	0			0
0	0	0	0	0	0					0	0	0			0
0	0	0	0	0	0					0	0	0			0
27622	1650	300	8770		645					5176	150	2693			
38342				33434	5971							2693			80440

EXCAVATION								STRUCTURE	GENERAL						
ACCEPTABLE			UNACCEPTABLE			Total Other than Class 5A	E.O. Hard Material		OTHER THAN 1C OR 6B			SPECIFIED 1C		SPECIFIED 6B	
3	Other than Class 3 and Class 5A	Total Other than Class 5A	U1	U2	Total				Above Structural Foundations	Fill to Structures	On Bridges	Above Structural Foundations	Fill to Structures	Above Structural Foundations	Fill to Structures
1	2	3	4	5	6	7	8		9	10	11	12	13	14	15
	110	110	50		50	160	10	SOUTH LANE OVERBRIDGE	500		50				
	250	250	75		75	325	0	STRAIGHT STREET OVERBRIDGE	554		60				
			150		150	150	0	REINFORCED EARTH STRUCTURE AT CH 3540-3600							
			210		210	210	0	CORRUGATED STEEL STRUCTURE - RIDGE VALE							
	360		485					Sub-Totals	1054		110				
360		360	485		485	845	10	TOTALS	1164						

FILL																		
SELECTED GRANULAR											SELECTED COHESIVE							Total Fill
6H		6I		6J		6K	6L	6M	6N/6P		7A		7B			7C	7D	
Drainage Layer to Reinforced Earth	Drainage Layer to Anchored Earth	Reinforced Earth	Anchored Earth	Reinforced Earth	Anchored Earth	Lower Bedding to Corrugated Steel Structures	Upper Bedding to Corrugated Steel Structures	Surround to Corrugated Steel Structures	Above Structural Foundations	Fill to Structures	Above Structural Foundations	Fill to Structures	Above Structural Foundations	Fill to Structures	Reinforced Earth	Reinforced Earth	Reinforced Earth	
16	17	18	19	20	21	22	23	24	25	26	27	28	29	30	31	32	33	34
										2850								3400
										3110								3724
120		3200																3320
						200	50	3000										3250
120		3200								5960								13694
120		3200				250	50	3000	5960